Crop Protection Under Changing Climate

Khawar Jabran • Singarayer Florentine
Bhagirath Singh Chauhan
Editors

Crop Protection Under Changing Climate

Springer

Editors
Khawar Jabran
Department of Plant Production and Technologies, Faculty of Agricultural Sciences and Technologies
Nigde Omer Halisdemir University
Nigde, Turkey

Singarayer Florentine
Centre for Environmental Management
Federation University Australia
Mt Helen, Ballarat, VIC, Australia

Bhagirath Singh Chauhan
The Centre for Crop Science
Queensland Alliance for
Agriculture and Food Innovation (QAAFI)
The University of Queensland
Gatton, QLD, Australia

ISBN 978-3-030-46110-2 ISBN 978-3-030-46111-9 (eBook)
https://doi.org/10.1007/978-3-030-46111-9

This Springer imprint is published by the registered company Springer Nature Switzerland AG
The registered company address is: Gewerbestrasse 11, 6330 Cham, Switzerland

Preface

Climate change is a result of anthropogenic activities, and subsequently has impacted all life on Earth. Scientists focus on the impacts of climatic changes on humans, animals, natural ecosystems, and agricultural systems. One important component of the impact of climate change on agricultural systems is the effect on crop pests. This book discusses the effects of two important components of climate change (global warming and rising carbon dioxide levels in the atmosphere) on plant pests, particularly weeds, insect pests and plant disease pathogens (including fungi and bacteria).

The chapter "Agriculture and Crop Protection; Its Global Importance and Relationship with Climate Change" explains the current state of climate change and its relationship with agriculture and crop protection, and the chapter "Assuring Crop Protection in the Face of Climate Change Through an Understanding of Herbicide Metabolisms and Enhanced Weed Control Strategies" addresses the changes in the behaviour of herbicides in response to elevated carbon dioxide in the levels of atmosphere and suggests strategies for weed control under climate change. The chapter "Climate change and Weeds of Cropping Systems", "Climate Change: Impact on Plant Pathogens, Diseases, and then suggests management" and "Potential for an Impact of Global Climate Change on Insect Herbivory in Cereal Crops" elaborate the interactions of climate change with weeds, disease pathogens and insect pests, respectively, and then suggests management strategies for these pests under climate change. The chapter "Crop Protection Under Drought Stress" and "Impact of Climate Change on Crop Production: Effects and Management" narrate the relationship between drought and crop production, and the impact of climate change on crop yields, respectively. The final chapter "Impacts of Climate

Change on Weeds, Insect Pests, Plant Diseases and Crop Yields: Synthesis" is a summary of the book and provides a synthesis of the book's contents.

Nigde, Turkey Khawar Jabran
Mt Helen, Ballarat, VIC, Australia Singarayer Florentine
Gatton, QLD, Australia Bhagirath Singh Chauhan

Contents

Agriculture and Crop Protection; Its Global Importance and Relationship with Climate Change ... 1
Singarayer Florentine, Bhagirath Singh Chauhan, and Khawar Jabran

Assuring Crop Protection in the Face of Climate Change Through an Understanding of Herbicide Metabolisms and Enhanced Weed Control Strategies ... 17
Simerjeet Kaur, Khawar Jabran, Singarayer Florentine, and Bhagirath S. Chauhan

Climate Change and Weeds of Cropping Systems ... 57
Annie Ruttledge and Bhagirath S. Chauhan

Climate Change: Impact on Plant Pathogens, Diseases, and Their Management ... 85
Mandeep Singh Hunjan and Jagjeet Singh Lore

Potential for an Impact of Global Climate Change on Insect Herbivory in Cereal Crops ... 101
Finbarr G. Horgan

Crop Protection Under Drought Stress ... 145
Ram L. Ray, Peter A. Y. Ampim, and Ming Gao

Impact of Climate Change on Crop Production: Effects and Management ... 171
Azeem Tariq and Muhammad Adil Rashid

Impacts of Climate Change on Weeds, Insect Pests, Plant Diseases and Crop Yields: Synthesis ... 189
Khawar Jabran, Singarayer Florentine, and Bhagirath Singh Chauhan

Index ... 197

Contributors

Peter A. Y. Ampim Cooperative Agricultural Research Center, College of Agriculture and Human Sciences, Prairie View A&M University, Prairie View, TX, USA

Bhagirath Singh Chauhan The Centre for Crop Science, Queensland Alliance for Agriculture and Food Innovation (QAAFI), The University of Queensland, Gatton, QLD, Australia

Singarayer Florentine Centre for Environmental Management, Federation University Australia, Mt Helen, Ballarat, VIC, Australia

Ming Gao Cooperative Agricultural Research Center, College of Agriculture and Human Sciences, Prairie View A&M University, Prairie View, TX, USA

Finbarr G. Horgan EcoLaVerna Integral Restoration Ecology, Bridestown, Kildinan, Co. Cork, Ireland

Mandeep Singh Hunjan Punjab Agricultural University, Ludhiana, India

Khawar Jabran Department of Plant Production and Technologies, Faculty of Agricultural Sciences and Technologies, Nigde Omer Halisdemir University, Nigde, Turkey

Simerjeet Kaur Punjab Agricultural University, Ludhiana, Punjab, India

Jagjeet Singh Lore Punjab Agricultural University, Ludhiana, India

Muhammad Adil Rashid Department of Plant and Environmental Sciences, University of Copenhagen, Copenhagen, Denmark

Ram L. Ray Cooperative Agricultural Research Center, College of Agriculture and Human Sciences, Prairie View A&M University, Prairie View, TX, USA

Annie Ruttledge Department of Agriculture and Fisheries, Toowoomba, QLD, Australia

Azeem Tariq Department of Plant and Environmental Sciences, University of Copenhagen, Copenhagen, Denmark

Agriculture and Crop Protection; Its Global Importance and Relationship with Climate Change

Singarayer Florentine, Bhagirath Singh Chauhan, and Khawar Jabran

Introduction

Agricultural practice, which includes well-established systems of cropping, pasture and forestry, represents a continual and essential dependence on healthy arable land across the globe and requires safeguarding with sustainable fertilization and pest control measures. This natural resource system must be continually protected from deliberate and inadvertent damage, in order to provide a suitable source of current and future amenities for all inhabitants of the planet. In this respect, it is morally and ethically necessary that we strive to manage the productivity and well-being of agricultural land in a way that will fulfil the necessities of the present generations and do not compromise needs of the future generations (Bruntland et al. 2012).

To put the magnitude of this issue in perspective, it is estimated that approximately 1.5 billion ha of fertile area is being used *for cropping purposes alone* and that 49% of the world's food requirements now come from this crop production activity (Tramberend et al. 2019). Because of this increasing reliance on crops as staple foodstuffs and because of its historical propensity for periodic failure (Mendelsohn 2007), this work will be specially focused upon the needs and problems experienced in both maintaining and expanding the quantity and quality of

S. Florentine
Centre for Environmental Management, Federation University Australia,
Mt Helen, Ballarat, VIC, Australia

B. S. Chauhan (✉)
The Centre for Crop Science, Queensland Alliance for Agriculture and Food Innovation (QAAFI), The University of Queensland, Gatton, QLD, Australia
e-mail: b.chauhan@uq.edu.au

K. Jabran
Department of Plant Production and Technologies, Faculty of Agricultural Sciences and Technologies, Nigde Omer Halisdemir University, Nigde, Turkey

K. Jabran et al. (eds.), *Crop Protection Under Changing Climate*,
https://doi.org/10.1007/978-3-030-46111-9_1

crop yields. In particular, it is noted that a characteristic of grain crops is that they are particularly sensitive to variation in growth conditions and attacks by aggressive pest species, and therefore crops require continual monitoring and attention to prevent poor harvests and, in extreme cases, from catalyzing widespread famine conditions (Anderson et al. 2019; Gaupp et al. 2020).

In order to achieve the productive and protective balance required of a sustainable crop situation, there are two key concerns which must be constantly addressed. First, studies have shown that the global population will continue to increase and will most likely reach 9.7 billion towards the middle of this century (UN DESA 2019). Although diets obviously vary widely across the world in terms of quantity, quality and constituent materials, it is clear that as a result of this expanding demand, to maintain reasonable and healthy levels of food consumption, there will be increasing pressure for production of all types of edible foodstuffs, including seafood, vegetables, various meats and, in particular, staple crops. This situation will become increasingly acute even though research and development in agricultural industries have been continually developing innovative and novel approaches to food production and diet change (Ciceri and Allanore 2019; Guerrero et al. 2019; Kristinsson and Jörundsdóttir 2019; Ruben et al. 2019). An important implication here is that fertile agricultural land will become under increasing production pressure and will require constant replenishment of minerals and organic material in order to ensure quality levels of plants are maintained.

A second concern, which has more recently become apparent, is that of increasingly severe and variable weather conditions, in conjunction with raised levels of atmospheric CO_2. The combined effects of this phenomenon, known as climate change, are already beginning to evidence unanticipated pressures on the food production industry. These pressures have been identified as having two main foci: the *direct effect* of raised CO_2 levels on plant biochemistry and the *indirect effects* on insect pests and weed species which impact on plant production. It has been noted (Elias et al. 2019; Lehmann et al. 2020) that this is an extremely complex area of concern because the flora and fauna involved in reaction to climate change are systemically linked with the condition of the soil, making reductive studies of this issue increasingly problematic. This complexity suggests that attempts at control of external pressures on crops, or restoration of soil in previously affected areas, should thus be approached with some caution.

This second concern, particularly when viewed through the lenses of crop production, protection and management, will be the special focus of this section. In order to begin to put this pressing issue of crop protection, in the face of climate variations, in a workable context, a brief review of pertinent literature is given here in order to establish what is currently known about the actual physical and biochemical changes catalyzed by climate change. This book chapter has been provided as a foundation for the following sections which report on current investigations into the practical issues involved with (i) the control and management of agronomic crop weeds and (ii) the implications of potential insect pest species expansion on crop production.

The Physical and Biochemical Manifestations of Global Climate Change

Prior to the industrial revolution, it has been estimated that, for about 1000 years, atmospheric CO_2 concentration was relatively stable at about 270 μmol mol^{-1} (Leakey et al. 2009). However, with the rapid increase in industrialization, CO_2 levels began accumulating at an alarming rate in the global atmosphere, and in the last decade, a growing number of scientific papers have considered the broad effects of climate change on a range of agricultural sectors across the globe. It is now widely recognized that global climate change will be growing as a problem into the foreseeable future, with atmospheric carbon dioxide concentrations predicted to rise to between 540 and 970 ppm over the next century (Prentice et al. 2001).

Of particular concern here is that it has been well documented (Thomson et al. 2010; Battisti and Larsson 2015; Ziska et al. 2019; Sharma and Dhillon 2020) that climate change and the damaging activities of pest species (both weeds and insects) are linked, and in this brief chapter, we will provide some essential descriptions of the basic effects of climatic changes which might influence pest distribution and expansion. In later sections, the chapter will look more closely at (i) the effects on weed biology, growth and control, particularly those found within crop plantations, and (ii) change in insect activities, together with their altered distribution and expansion, in the vicinity of crop production.

Direct Effects of Increased Atmospheric CO_2 Levels

It has been established that these increases in atmospheric CO_2 are directly responsible for changes to atmospheric temperature and ocean acidity, both of which cause a range of complex interacting flow-on effects to climate, weather and ocean chemistry (Houghton 2001). It has been postulated that this will stimulate the increase of mean global temperature, leading to the stimulation of climate change events including more frequent drought and heat waves (Mpelasoka et al. 2008). Indeed, the International Panel for Climate Change reports that the global average surface temperature will continue to increase by a further 1.4–5.8 °C by 2100 (IPCC 2013). As a consequence, much more extreme natural events such as fire, flooding and drought will be expected in Australia (Bureau of Meteorology 2014), which will have significant implications for the agricultural industry. It has also been established that, other than atmospheric CO_2 increase, drought, salinity and extreme temperatures are the most important environmental factors that affect the healthy growth of plants.[1] Current efforts at the prediction of the effects on climate and weather

[1] The increase in stratospheric ozone (O_3) is another concerning key factor in future atmospheric changes as it can cause visual plant injuries (Benton et al. 2000). The thinning in protective ozone

through the modelling of future climate scenarios (Najafi and Kermani 2017; Fernandez et al. 2019) indicate that temperature change is more likely to be accurately provided than changes to rainfall patterns. It is likely that, in the future, the number and magnitude of extreme weather events will increase in frequency and severity and will continue to be highly unpredictable (Hughes 2003). Given this situation, investigating the responses evidenced by weeds and pest insects to extreme weather conditions (flood and drought and heat and cold) is likely of importance to those charged with the management of the impact of climate change on Australian agriculture. Such investigations must be of high priority, particularly in the context of crop production concerns, to ensure continuity of supply.

Reactions of Plants to Increased CO_2 Levels

As a first consideration of the ways in which living organisms will respond to climate change, it has been opined that plants are likely to have greater influence on insects than vice versa (Dukes and Mooney 1999); thus we have taken a discussion of the possible effect of climate change on plants as an appropriate starting point.

It has been observed (Warren et al. 2018; Schleuning et al. 2020) that climate change effects will act in complex ways on plants, and this will affect plant fitness at a species level as well as setting the environmental cues for successful germination, growth, maturity, seed set and seed dispersal. The direct effects on plant physiology of elevated CO_2 levels include increased plant biomass, reduced water uptake from the soil, alteration in the C-N ratios found in plant tissues and increased resistance to herbicides (Ziska and Bunce 1997; Ward et al. 1999; Patterson 1995; Wand et al. 1999; Nicotra et al. 2010; Walther et al. 2009; Mahajan et al. 2012). This is, of course, occurring in parallel with the indirect effects of climate change, which is the influence of elevated atmospheric CO_2 on atmospheric and oceanic temperatures and the consequent changes in rainfall and wind patterns (Hughes 2003). This will contribute to the complexity of understanding the potential effects on plant development since the combined result of endogenous and exogenous pressures on plant development needs to be considered.

Another important element in this equation is plant type. There are at two different metabolic plant types, divided according to their mode of carbon metabolism, and these are known as C_3 or C_4 types. It has been found that each of these types is adapted to different climatic conditions; C_3 species are termed cool season plants, and C_4 species are called warm season plants (Pearcy and Ehleringer 1984). In Australia, C_3 and C_4 grass species, for example, are distributed according to climate, with C_3 types found in regions with a cool, wet spring and C_4 types in areas with a hot, wet summer (Hattersley 1983). In stable weather conditions, therefore, it would

layer also can lead to increased levels of ultraviolet (UV) radiations causing implications for crop cultivars and plants (Krupa et al. 1998).

be expected that plants with the same carbon metabolism will flourish in separate regions, but any significant alterations in these climatic zones will begin to influence changes in species distribution.

In this respect, the southern region of Australia has been typically Mediterranean in character, comprising cool, wet winters and spring and warm, dry summers. By contrast, the northern region of the continent is mostly tropical, with hot, wet summers and cooler, dry winters. However, changes in climate driven by mechanisms known as El Niño and La Niña (Philander 1985), and as recorded by the Southern Oscillation Index and Indian Ocean Dipole patterns (Saji and Yamagata 2003), appear to be signalling changes to long-term weather patterns in Australia. These changes are manifest as an increase in the magnitude and frequency of heavy rainfall events, as well as longer and deeper drought conditions. This evidence of climate change is, therefore, presaging a potentially damaging interference with the stability of long-term climate trends in Australia (Hughes 2003). Consequently, the traditional distributions of C_3 and C_4 plants, including agricultural weeds, may become significantly altered over time due to this variability and thus have considerable implications for future agricultural management.

Effects of Climatic Changes on Weed Distribution, Biology, Growth and Control

As commented earlier, the overall effects on plants from these anticipated changes in climate are a result of a number of interacting responses. Changes to temperature and moisture availability act as stressors for all plants and are due to changes in seasonal temperature and precipitation event variations across large geographic regions. Of particular concern to agricultural practice is that these climatic changes subsequently result in the changed distribution patterns of weeds as they react to their physiological responses to these external stresses, and, at the same time, this manifestation of genetic fitness may enable them to begin to aggressively adapt to the new conditions (Nicotra et al. 2010; Clements and Ditommaso 2011; Gallagher et al. 2013).

As a consequence of this changing climatic situation, alterations in the distribution of weeds across tropical and temperate regions will ultimately be determined by a combination of climatic effects and geographic factors. In the Northern Hemisphere,[2] there has been a trend for polewards movement of some plant populations with increasing temperatures (Walther et al. 2009), and it is anticipated that this shift is also likely to occur in Australia. Traditional 'tropical' weeds can be expected to expand their ranges southwards as will also be the case with warm-

[2] For example, weeds such as itchgrass (*Rottboellia cochinchinensis*), which is currently restricted to the Mediterranean region, with its tropic and sub tropic conditions, are expected to expand their range to temperate regions with the expected future climatic conditions (Fuhrer 2003).

adapted temperate species, but, conversely, the ranges of cool-adapted temperate species will likely be reduced (Gallagher et al. 2013). It has been suggested (Ref) that some species of weeds may become extinct under these change conditions since the Australian continent does not cover the same extreme latitudinal range as Northern Hemisphere land masses. This will consequently deny some cold-adapted species sufficient area of refuge for survival, and, in addition, distribution changes and survival of weeds will also be challenged by the availability of suitable soil types for each particular species. Factors such as soil pH and nutrient availability are strong drivers for the ability of weeds to become established and to thrive, and if these conditions are not appropriate or limited in extent, despite other factors of temperature and moisture availability being suitable, changes in the range of such weeds may be unlikely to occur. It might be commented that plants grown under predicted future climate changes have shown a variety of phenotypic differences (Edwards et al. 2010; Kriticos et al. 2010). One of the most obvious changes in plants under elevated CO_2 levels is the increase in above and/or below ground biomass, higher photosynthetic output and increased seed production (Long et al. 2004) which is typical of invasive weed species.

Notwithstanding the negative effects of these stressors on weed species, because of the adaptability and development of their biochemical and physiological strategies, it has been widely observed that weed species can become remarkably well adapted to environmental stress conditions (Patterson and Flint 1990). These adaptive strategies, which give weeds their characteristic abilities for re-establishment in widely varying agricultural situations, are based on either avoidance of potentially damaging factors and/or the acclimatization of the plant's system to the altered conditions. They manage this by offsetting the novel biochemical and physiological conditions by such modes as the accumulation of antioxidants, osmoprotectants and heat shock proteins, together with signalling cascades and transcriptional control (Hasanuzzaman et al. 2013). The success of these strategies is clearly evident, as it is well-known that weed species usually outcompete crop plants in terms of the use of available resources, even under changing and challenging climatic conditions. In terms of the focus of this discussion, it has been specifically observed that the extent of this competition between crop and weed species particularly depends on the level of atmospheric CO_2 (Patterson and Flint 1990).

In this latter respect, it has been observed that, depending upon the plant's functional group, yield response and degree of plant growth vary with CO_2 availability (Taub et al. 2008); at the same time, plant growth rate also depends on the ambient temperature of the air, available soil nutrients and water interactions (Kimball et al. 2002). This multifaceted situation implies that not all plant species are evenly benefited by increases in atmospheric CO_2, but, in the majority of cases, weeds evidence greater advantages than crop plants. These reactive abilities are likely to make weeds an increasing challenge for economic crop cultivation under predicted climatic conditions in the future. It has also been clearly demonstrated that aggressive weeds can indirectly determine the crop yield potential of a crop plant, as many the weed species are excellent in adapting to changing environmental conditions, including stresses such as water deficit and high temperatures (Patterson and Flint 1990).

Functional Differences in C_3 and C_4 Plants

It has also been established that, under changing climatic conditions, C_3 and C_4 plants differ in their competitive adaptability. This is thought to be due to the differences in the biochemical processes involved in their photosynthetic pathways. Whilst under elevated atmospheric CO_2 levels, the water use efficiency of both C_3 and C_4 plants is enhanced due to reduced stomatal conductance (Ainsworth and Long 2005), an added advantage for C_3 plants is that they are capable of enhancing their photosynthetic capacity, a feature of which C_4 plants are not capable. Further, evidence suggests that, in the absence of other stressful conditions, many C_3 species have evidenced a strong response to increased atmospheric CO_2 (Leakey et al. 2009; Long et al. 2004; Ziska 2000; Ziska and Bunce 1997). Investigations have determined that in C_4 plants, the CO_2 fixation mechanism becomes saturated at a CO_2 concentration of 360 ppm, leaving no leeway for further CO_2 fixation (Leegood, 2002). As 95% of the world crop plants belong to the C_3 photosynthesis pathway, they are potentially more advantaged when competing with C_4 weeds under elevated atmospheric CO_2.

In the study by which was carried out under elevated atmospheric CO_2 levels, an increase in biomass has been observed in both C_3 and C_4 species. In a similar investigation, according to Fernando et al. (2017), *Lolium rigidum* (annual ryegrass) responded to conditions of elevated CO_2 with increased total biomass production, clearly demonstrating its competitiveness in these conditions. Increased biomass of plants under elevated CO_2 has also been evidenced by in a long-term free-air CO_2 enrichment (FACE) experiment involving *Toxicodendron radicans* (poison ivy), implying an increase in rate of growth and biomass, photosynthesis and water use efficiency under enhanced atmospheric CO_2 levels. A significant study by Ziska (2000) has shown that either C_3 or C_4 weeds which have infested soybean crops could reduce the soybean yield under both ambient and elevated CO_2 conditions. This study has also revealed that the ability of soybean crops to respond positively to growth conditions has been reduced under elevated CO_2, especially when competing with C_3 weeds. This has been suggested to be due to the success of weeds to aggressively compete for available resources including nutrients, water and minerals.

However, it must be noted that, in contrast, elevated CO_2 concentration can also have a negative impact on the growth and biomass of some species. According to the study of Hovenden et al. (2008), the quality of seed, its weight, germination and seedling growth of *Austrodanthonia caespitosa* (wallaby grass) have been significantly sensitive to increased CO_2 levels (550 ppm) and warming (140 W m^{-2}), with reduced parameters under elevated CO_2 conditions. Warming has also reduced the seed germination success and subsequent seedling growth. It is also somewhat surprising to note that C_4 species within the prairie plant communities have unexpectedly responded more positively to elevated CO_2 concentrations than C_3 species. In this situation, the changes in the biomass and seed production are mainly driven by nitrogen availability.

With these confounding observations in mind, it is clearly necessary to consider management strategies as being subject to case-specific reactions to climate change. Nonetheless, there are some tentative approaches which are emerging. Adjustments in resource management could be one key aspect in facilitating the optimal conditions. For example, careful adaptation of irrigation techniques to cope with changing precipitation patterns in order to preserve optimum soil conditions is essential. Also, approaches to managing nutrient availability will be important, as the increasing atmospheric CO_2 levels and air temperature can influence nitrogen cycling. It has been pointed out that advanced genetic breeding and screening methods will also be advantageous in producing high yielding and competitive crop types to compete with weeds, which, in general, can outcompete most crop species (Kobiljski and Dencic 2001).

Increasing Herbicide Tolerance Issues

It has also been suggested that herbicide tolerance in weeds might be affected by changes in future climatic conditions (Fuhrer, 2003). In a recent study, the C_3 weed species, Lamb's quarters (*Chenopodium album* L.), has shown to be more herbicide-tolerant when treated with glyphosate under elevated CO_2 levels than when it is applied under ambient conditions (Ziska and Teasdale 2000; Ziska et al. 1999). This suggests that careful herbicide application strategies should also be implemented in cropping systems when dealing with weed control under expected future climatic changes. Unless cautious attention is given to intensive agriculture and management, the problems associated with weeds are expected to become greater in future, with weeds becoming progressively immune to herbicide control.

Some Comments on Predictive Modelling

It has been recommended that the use of predictive modelling, such as CLIMEX which can characterize the weeds that may increase their ranges into particular cropping regions, will assist in designing management actions to address new outbreaks of particular weeds if and when they are detected (Wilson et al. 2009, Gallagher et al. 2013). However, unless such models include factors that are not directly climate related, such as soil type, or else if preferences for such factors are unknown in the species of weeds being modelled, their future distributions cannot be accurately predicted. Genetic variability of weeds is another factor that could cause inaccuracies of modelling future weed distributions under climate change (Clements and Ditommaso 2011). Genetic makeup is fundamental for the responses to environmental stimuli at individual and population levels of these organisms, and genetic plasticity will enable some weeds to respond more effectively to climate change than others.

Reactions of Insects to Increased CO_2 Levels

Although changes in the distribution of host plants of pest insects will significantly influence insect distribution under future climate change, insects may also respond independently according to factors such as temperature and rainfall. However, it is agreed that they will be less directly affected by the amount of atmospheric carbon dioxide than plants. As with changes in the distribution of weeds, most research on the effect of climate change on insects has been conducted in the Northern Hemisphere. For example, observations of butterflies, moths and ants have been recorded and correlated with increased temperatures and changes to rainfall patterns in Europe and North America (Walther et al. 2002, 2009). However, it seems that the findings are broadly applicable to Australia (Andrew et al. 2013) since the expected changes in climate and weather patterns are global.

Insects are useful indicators of human-driven climate change, due to the observable effects in their distribution, phenology, population structure, and evolutionary changes at population levels (Menendez 2007). Like all other living organisms, each insect species is adapted to a particular set of climatic conditions. Their life cycle cues are intimately linked to seasonal patterns, including, for example, variations in photoperiod and the occurrence of low temperatures, which may induce periods of diapause, and elevated temperature thresholds for activity such as mating and egg laying. Disruptions to these processes can be expected to have profound effects, but may not necessarily be disadvantageous in all cases. For example, elevated seasonal temperatures can drive rapid population growth by increasing the number of generations per year or season (Rosenzweig et al. 2001; Thomson et al. 2010; Sutherst et al. 2011), and changes in the timing of breeding may allow other species to avoid predation by natural enemies (Harrington et al. 2001; Rosenzweig et al. 2001; Fuhrer 2003; Thomson et al. 2010). Alternatively, increases in drought may reduce the effectiveness of fungi as biological control organisms of insects (Sutherst et al. 2011), and other such negative impacts may be a reduction of effectiveness of insect biological controls, due to changes in emergence times of their prey species (Thomson et al. 2010).

To illustrate the possibilities for the impact of insect pests on Australian agriculture, it is useful to find examples that are a result of possible climate change effects from the Northern Hemisphere. Such an example from the United States is the observation that major pest outbreaks appear to be strongly correlated with El Niño and La Niña events. For instance, extensive damage to soybean crops from two-spotted mite, *Tetranychus urticae*, occurs during drought, and similar major outbreaks of insect pests have been observed to occur during more favourable (to crops) weather conditions (Rosenzweig et al. 2001).

Other effects of a warming climate include (i) the overwinter survival of larger numbers of insect pests or (ii) their earlier seasonal emergence from hibernation following cooler times of the year. Larger early season numbers of pest insects may significantly reduce their host plant species' survival, due to alterations in the timing of insect attacks occurring during more vulnerable life cycle stages of the plants.

However, subsequent higher temperatures during summers may also increase mortality or pose other challenges for breeding of insects (Sutherst et al. 2011). It can be reasonably expected that different insect species, but which have similar preferences for crops in Australia that are grown elsewhere, will respond in the same manner as those in the Northern Hemisphere to climate change.

For example, the Queensland fruit fly is expected to move from tropical regions to more southerly parts of the continent as seasonal temperatures and rainfall patterns change to become more similar to its current range (Sutherst et al. 2011). This is expected to impact not only agricultural production but also backyard growers (Sutherst 2000). Unless appropriate management strategies are available, which include, first and foremost, biosecurity to prevent incursions, this insect will cause significant economic damage to the fruit growing industry of southern Australia.

The threats by pest insects to agricultural systems are predicted to continue under the predicted future climatic changes, but this issue has been rarely taken into account compared to the significant focus on the importance of weeds. However, future climatic changes can affect insects in both direct and indirect ways as they can influence environmental conditions, chemical composition, host physiology, predators, pathogens and competitiveness (Patterson 1995). With atmospheric warming, insect pests in the mid- to high-latitude regions are expected to significantly increase in number in the future, since temperature plays a major role in the insect life cycle, and thus climate warming is expected to directly affect their life cycles (Fuhrer 2003). Indeed, it has been claimed that, in the temperate zone, it is expected that insect winter survival will increase, and at the northern latitudes, phenological shifts including growth and reproduction will be the main effects of climate warming (Bale et al. 2002).

The geographical range of many insect populations is also expected to be changed due to changes in climatic conditions. However, expansion of range is somewhat dependent on the host's ability to support an invasion of new species (Fuhrer 2003). It is suggested that migrant pests are able to respond more quickly than native species, as they can quickly colonize the existing crops in distant places (Cannon 1998). Other authors have indicated that the host plants may get affected by the predicted future drought conditions, making them more attractive to pest species; excessive precipitation, on the other hand, may lead to water logging conditions affecting the soil dwelling insects (Watt and Leather 1988). Indirect effects can also arise and can involve insect pathogens, predators and parasites (Raulston et al. 1992). In Australia, as the insect population size and quality are highly dependent on the high variability in rainfall, it is anticipated that drastic changes in such aspects are expected as the rainfall patterns change over time. (Drake 1994). For insect outbreaks in places far from their normal range, extreme triggers such as droughts and floods are important. Therefore it is expected that the shifts and increased frequency of these triggers will affect many agricultural ecosystems.

As a result of elevated CO_2 levels, it is predicted that the plant chemical composition, including C-N ratio, allelochemical concentrations and starch and fibre content may be altered (Lincoln et al. 1993). The reduced N and elevated carbohydrate concentration in plants under elevated CO_2 levels make them preferred diets for insects, as they are more easily digestible (Goverde et al. 1999). However under such conditions, the leaf chewers are suggested to compensate their low N intake by increasing the quantity of food ingested, whilst leaf miners are affected to a lesser degree and seedeaters being not affected at all (Bezemer and Jones 1998). As the elevated CO_2 levels can increase the thickness of plant leaf epidermis and specific leaf weight, the response of leaf-sucking insects, including mites, is expected to be negatively affected (Joutei et al. 2000), but as suggested by Heagle et al. (2002), elevated CO_2 levels can increase the leaf temperature, positively affecting the fecundity of insects.

In summary, based on this survey of recent studies, it appears that the impact of climate change on insect pests is expected to include issues such as increase in population size and growth rate; facilitated overwintering ability; an upturn in the number of generations, distribution range and migration habits; and expanded host-pest synchronization. However, all the above changes will largely depend on the species genotypic and phenotypic flexibility. The effect of temperature on insects is also dependent on the species, their ability to adapt to change and also their life histories (Bale et al. 2002). It is clear that long-term predictions have become increasingly difficult as the understandings of interactions among ecosystem processes, human influences and environmental conditions, which play a major role, continue to emerge.

Section Two: Agronomic Crop Weeds' Responses to Climate Change

As indicated earlier, increased levels of atmospheric carbon dioxide are likely to have both direct and indirect effects on plants, including a range of agricultural weeds related to crop production upon which this work is focused. In the first instance, it is noted that the likely observable responses of weeds to climate change are faster growth and earlier maturation than the crops they infest (Ziska 2016), increased changes to seed longevity (Nicotra et al. 2010), increased overall biomass (Ziska and Bunce 1997; Wand et al. 1999; Ward et al. 1999, Ziska 2010) and increased root biomass (Mahajan et al. 2012, Bansal et al. 2014). All of these alterations allow increased competition advantages for weeds over crops for moisture and nutrients (Patterson 1995; Wand et al. 1999; Ward et al. 1999).

Section Three: Implications for Crop Protection due to Pest Species Expansion Stimulated by Conditions Arising from Climate Change Factors

It is asserted here that although the impacts of anthropogenic climate change have been observed in changes to the geographic distribution and life cycles of plants and animals in many regions of the world (Hughes 2003), globally, modelling for future scenarios of these impacts on life forms is at a relatively early stage and is not yet well defined for more than a few species. In particular, we note that in Australia this work is at a relatively early stage, but it is relevant to point out that multiple parameter models are beginning to be employed for making predictions about selected species (Andrew et al. 2013).

Effects of Climatic Changes on Pest Insect Activities, Their Distribution and Expansion

There is a high potential of multiple changes in the activities and distribution of pest insects, including population densities, timing of life cycle events and feeding behaviour (Pelini et al. 2009; Getahun 2020). These may occur when already established pest species undergo adaptation in situ and cause more damage to crops or as a result of temperature and rainfall changes opening up niches in new locations (Battisti and Larsson 2015). Additionally, more rapid dispersal of pest species newly introduced to some regions may also occur because of an increase in the number extreme weather events, such as hurricanes (Hellmann et al. 2008). More rapid reproductive rates of species that can take advantage of increased resources or an absence of low temperatures that would normally induce a reproductive diapause may also be expected under a scenario of increased year-round temperatures. Therefore, pest species that typically undergo an overwinter diapause may have opportunities to breed all year round, potentially increasing crop damage to previously unseen scales (Pareek et al. 2017). Such events will require management agencies to respond more urgently than previously required, in order to minimize crop damage. These climatic variables are also unlikely to operate as the sole potential drivers for changes in distribution, population and life cycles of pest insects, however. Changes in plant biology, phenology and distribution in response to changes of atmospheric carbon dioxide levels, temperature and rainfall patterns will also influence pest insect behaviour and distribution (Chander et al. 2016). However, not all changes may be beneficial for the survival of some species. The life cycle traits of species closely attuned to particular temperature ranges and rainfall patterns may be negatively impacted under significantly altered cycles of drought and above average rainfall (Yan et al. 2017). For example, some generalist insect species' survival may be reduced when either extreme drought or flood events occur at life cycle stages during which the insect population is relatively vulnerable, such as during the larval or pupal stage (Rai et al. 2018). Additionally, alterations in climate variables impacts may also affect beneficial insects, such as honeybees (Getahun ibid.), as well as insects used as biocontrols (Thomson et al. 2010).

References

Ainsworth EA, Long SP (2005) What have we learned from 15 years of free-air CO_2 enrichment (FACE)? A meta-analytic review of the responses of photosynthesis, canopy properties and plant production to rising CO_2. New Phytol 165(2):351–372

Anderson WB, Seager R, Baethgen W, Cane M, You L (2019) Synchronous crop failures and climate-forced production variability. Science Adv 5(7):eaaw1976

Andrew NR, Hill SJ, Binns M, Bahar MH, Ridley EV, Jung M-P, Fyfe C, Yates M, Khusro M (2013) Assessing insect responses to climate change: what are we testing for? Where should we be heading? Peer J 1:11

Bale JS, Masters GJ, Hodkinson ID, Awmack C, Bezemer TM, Brown VK et al (2002) Herbivory in global climate change research: direct effects of rising temperature on insect herbivores. Glob Chang Biol 8(1):1–16

Bansal S, James JJ, Sheley RL (2014) The effects of precipitation and soil type on three invasive annual grasses in the western United States. J Arid Environ 104:38–42

Battisti A , Larsson S (2015) Climate change and insect pest distribution range. In: Climate change and insect pests. CABI, Wallingford, pp 1–15

Benton J, Fuhrer J, Gimeno B, Skärby L, Palmer-Brown D, Ball G et al (2000) An international cooperative programme indicates the widespread occurrence of ozone injury on crops. Agric Ecosyst Environ 78(1):19–30

Bezemer TM, Jones TH (1998) Plant-insect herbivore interactions in elevated atmospheric CO_2: quantitative analyses and guild effects. Oikos 212–222

Brundtland GH, Ehrlich P, Goldemberg J, Hansen J, Lovins A, Likens G, Lovelock J, Manabe S, May B, Mooney H, Robert KH (2012) Environment and development challenges: the imperative to act. The Asahi Glass Foundation, Tokyo

Bureau of Meteorology (2014) What is El Niño and what might it mean for Australia? Retrieved 29.10.2017, from Bureau of Meteorology. http://www.bom.gov.au/climate/updates/articles/a008-el-nino-and-australia.shtml

Cannon RJ (1998) The implications of predicted climate change for insect pests in the UK, with emphasis on non-indigenous species. Glob Chang Biol 4(7):785–796

Chander S, Husain M, Pal V (2016) Insect pest management in climate change. In: Chattopadhyay C, Prasad D (eds) Dynamics of crop protection and climate change. Studera Press, New Delhi, pp 115–130

Ciceri D, Allanore A (2019) Local fertilizers to achieve food self-sufficiency in Africa. Sci Total Environ 648:669–680

Clements DR, Ditommaso A (2011) Climate change and weed adaptation: can evolution of invasive plants lead to greater range expansion than forecasted? Weed Res 51:227–240

Drake V (1994) The influence of weather and climate on agriculturally important insects: an Australian perspective. Aust J Agric Res 45(3):487–509

Dukes JS, Mooney HA (1999) Does global change increase the success of biological invaders? Trends Ecol Evol 14:135–139

Edwards EJ, Osborne CP, Stramberg CAE, Smith SA (2010) The origins of C_4 grasslands: integrating evolutionary and ecosystem science. Science 328, 587–591

Elias EH, Flynn R, Idowu OJ, Reyes J, Sanogo S, Schutte BJ, Smith R, Steele C, Sutherland C (2019) Crop vulnerability to weather and climate risk: analysis of interacting systems and adaptation efficacy for sustainable crop production. Sustainability 11(23):6619

Fernando N, Florentine SK, Naiker M, Panozzoc J, Chauh BS (2019) Annual ryegrass (*Lolium rigidum Gaud*) competition altered wheat grain quality: A study under elevated atmospheric CO_2 levels and drought conditions. Food Chemistry. 276: 285–290

Fernández J, Frías MD, Cabos WD, Cofiño AS, Domínguez M, Fita L, Gaertner MA, García-Díez M, Gutiérrez JM, Jiménez-Guerrero P, Liguori G (2019) Consistency of climate change projections from multiple global and regional model intercomparison projects. Clim Dyn 52(1–2):1139–1156

Fuhrer J (2003) Agroecosystem responses to combinations of elevated CO_2, ozone, and global climate change. Agric Ecosyst Environ 97(1):1–20. https://doi.org/10.1016/S0167-8809(03)00125-7
Gallagher RV, Englert DD, O'Donnell J, Wilson PD, Downey PO, Hughs L, Leishman MR (2013) The grass may not always be greener: projected reductions in climate suitability for exotic grasses under future climates in Australia. Biol Invasions 15:961–975
Gaupp F, Hall J, Hochrainer-Stigler S, Dadson S (2020) Changing risks of simultaneous global breadbasket failure. Nat Clim Chang 10:54–57. https://doi.org/10.1038/s41558-019-0600-z
Getahun D (2020) Predictions of climate change on agricultural insect pests Vis-à-Vis food crop productivity: a critical review. Ethiop J Sci Sustain Dev 7(1):18–26
Goverde M, Bazin A, Shykoff J, Erhardt A (1999) Influence of leaf chemistry of Lotus corniculatus (Fabaceae) on larval development of Polyommatus icarus (Lepidoptera, Lycaenidae): effects of elevated CO_2 and plant genotype. Funct Ecol 13(6):801–810
Guerrero Lara L, Pereira LM, Ravera F, Jiménez-Aceituno A (2019) Flipping the tortilla: social-ecological innovations and traditional ecological knowledge for more sustainable agri-food systems in Spain. Sustainability 11(5):1222
Harrington R, Fleming RA, Woiwod IP (2001) Climate change impacts on insect management and conservation in temperate regions: can they be predicted? Agric For Entomol 3:233–240
Hasanuzzaman M, Nahar K, Alam M, Roychowdhury R, Fujita M (2013) Physiological, biochemical, and molecular mechanisms of heat stress tolerance in plants. Int J Mol Sci 14(5):9643
Hattersley PW (1983) The distribution of C_3 and C_4 grasses in Australia in relation to climate. Oecologia 57:113–128
Heagle A, Burns J, Fisher D, Miller J (2002) Effects of carbon dioxide enrichment on leaf chemistry and reproduction by twospotted spider mites (Acari: Tetranychidae) on white clover. Environ Entomol 31(4):594–601
Hellmann JJ, Byers JE, Bierwagen BG, Dukes JS (2008) Five potential consequences of climate change for invasive species. Conserv Biol 22(3):534–543
Houghton J (2001) The science of global warming. Interdiscip Sci Rev 26(4):247–257. https://doi.org/10.1179/isr.2001.26.4.247
Hovenden MJ, Wills KE, Chaplin RC, Jacqueline K, Schoor V, Williams AL, Osanai Sanai, Y Newton PD (2008) Warming and elevated CO_2 affect the relationship between seed mass, germinability and seedling growth in Austrodanthonia caespitosa, a dominant Australian grass. Global Change Biology, 14, 1633–1641
Hughes L (2003) Climate change in Australia: trends, projections and impacts. Austral Ecol 28
IPCC (2013) Summary for Policymakers. In: Climate change 2013: the physical science basis. Contribution of working group 1 to the Fifth Assessment report of the intergovernmental panel on climate change retrieved from Cambridge University Press, Cambridge/New York
Joutei AB, Roy J, Van Impe G, Lebrun P (2000) Effect of elevated CO_2 on the demography of a leaf-sucking mite feeding on bean. Oecologia 123(1):75–81
Kimball B, Kobayashi K, Bindi M (2002) Responses of agricultural crops to free-air CO_2enrichment. Adv Agron 77:293–368
Kobiljski B, Dencic S (2001) Global climate change. Challenge for breeding and seed production of major field crops. J Genet Breed 55(1):83–90
Kristinsson HG, Jörundsdóttir HÓ (2019) Food in the bioeconomy. Trends Food Sci Technol 84:4–6
Kriticos DJ, Crossman ND, Ota N, Scott JK (2010) Climate change and invasive plants in South Australia. Report for the South Australian Department of Water, Land and Biodiversity Conservation. CSIRO Climate Adaptation Flagship, Canberra, Australia. 92pp.
Krupa SV, Kickert RN, Jäger H-J (1998) Elevated ultraviolet (UV)-B radiation and agriculture. Springer
Leakey ADB, Ainsworth EA, Bernacchi CJ, Rogers A, Long SP, Ort DR (2009) Elevated CO_2 effects on plant carbon, nitrogen, and water relations: six important lessons from FACE. J Exp Bot 60(10):2859–2876. https://doi.org/10.1093/jxb/erp096

Leegood RC (2002) C_4 photosynthesis: principles of CO_2 concentration and prospects for its introduction into C_3 plants. J Exp Bot 53(369):581–590
Lehmann P, Ammunét T, Barton M, Battisti A, Eigenbrode SD, Jepsen JU, Kalinkat G, Neuvonen S, Niemelä P, Terblanche JS, Økland B (2020) Complex responses of global insect pests to climate warming. Front Ecol Environ. https://doi.org/10.1002/fee.2160
Lincoln DE, Fajer ED, Johnson RH (1993) Plant-insect herbivore interactions in elevated CO_2 environments. Trends Ecol Evol 8(2):64–68
Long SP, Ainsworth EA,Rogers A, Ort DR (2004) Rising atmospheric carbondioxide :plants FACE the future. Ann Rev Plant Biol 55, 591–628.
Mahajan G, Singh S, Chauhan BS (2012) Impact of climate change on weeds in the rice-wheat cropping system. Curr Sci 102:1254–1255
Mendelsohn R (2007) What causes crop failure? Clim Chang 81(1):61–70
Menéndez R (2007) How are insects responding to global warming? Tijdschrift voor Entomologie 150:355–365
Mpelasoka F, Hennessy K, Jones R, Bates B (2008) Comparison of suitable drought indices for climate change impacts assessment over Australia towards resource management. Int J Climatol 28(10):1283–1292
Najafi R, Kermani MRH (2017) Uncertainty modeling of statistical downscaling to assess climate change impacts on temperature and precipitation. Water Resour Manag 31(6):1843–1858
Nicotra AB, Atkin OK, Bonser SP, Davidson AM, Finnegan EJ, Mathesius U, Poot P, Purugganan MD, Richards CL, Valladares F, van Kleunen M (2010) Plant phenotypic plasticity in a changing climate. Trends Plant Sci 15:684–692
Pareek A, Meena BM, Sharma S, Tetarwal ML, Kalyan RK, Meena BL (2017) Impact of climate change on insect pests and their management strategies. Available at: http://www.researchgate.net/publication/32847622
Patterson DT (1995) Weeds in a changing climate. Weed Sci 43(4):685–700
Patterson DT, Flint EP (1990) Implications of increasing carbon dioxide and climate change for plant communities and competition in natural and managed ecosystems. In: Impact of carbon dioxide, trace gases, and climate change on global agriculture(impactofcarbond), pp 83–110
Pearcy RW, Ehleringer J (1984) Comparative ecophysiology of C3 and C4 plants. Plant Cell Environ 7(1):1–13
Pelini SL, Prior KM, Parker DJ, Dzurisin JD, Lindroth RL, Hellmann JJ (2009) Climate change and temporal and spatial mismatches in insect communities. In: Climate change. Elsevier, pp 215–231
Philander SGH (1985) El Niño and La Niña. J Atmos Sci 42(23):2652–2662
Prentice IC, Farquhar G, Fasham M, Goulden M, Heimann M, Jaramillo V et al (2001) The carbon cycle and atmospheric carbon dioxide. Cambridge University Press, Cambridge
Rai B, Klein AM, Walter J (2018) Chronic dryness and wetness and especially pulsed drought threaten a generalist arthropod herbivore. Oecologia 188(3):931–943
Raulston J, Pair S, Loera J, Cabanillas H (1992) Prepupal and pupal parasitism of Helicoverpa zea and Spodoptera frugiperda (Lepidoptera: Noctuidae) by Steinernema sp. in cornfields in the lower Rio Grande Valley. J Econ Entomol 85(5):1666–1670
Rosenzweig C, Iglesias A, Yang XB, Epstein PR, Chivian E (2001) Climate change and extreme weather events; implications for food production, plant diseases, and pests. Glob Chang Hum Health 2:90–104
Ruben R, Verhagen J, Plaisier C (2019) The challenge of food systems research: what difference does it make? Sustainability 11(1):171–185
Saji NH, Yamagata T (2003) Possible impacts of Indian Ocean dipole mode events on global climate. Clim Res 25(2):151–169
Schleuning M, Neuschulz EL, Albrecht J, Bender IM, Bowler DE, Dehling DM, Fritz SA, Hof C, Mueller T, Nowak L, Sorensen MC (2020) Trait-based assessments of climate-change impacts on interacting species. Trends Ecol Evol

Sharma HC, Dhillon MK (2020) Climate change effects on arthropod diversity and its implications for Pest management and sustainable crop production. Agroclimatol Linking Agric Climate 60: 595–619. Link to abstract: https://acsess.onlinelibrary.wiley.com/doi/abs/10.2134/agronmonogr60.2016.0019
Sutherst RW (2000) Climate change and invasive species: a conceptual framework. In: Mooney HA, Hobbs RJ (eds) Invasive species in a changing World. Island Press, Washington
Sutherst RW, Constable F, Finlay KJ, Harrington R, Luck J, Zalucki MP (2011) Adapting to crop pest and pathogen risks under a changing climate. Wiley Interdiscip Rev Clim Chang 2:220–237
Taub DR, Miller B, Allen H (2008) Effects of elevated CO_2 on the protein concentration of food crops: a meta-analysis. Glob Chang Biol 14(3):565–575
Thomson LJ, Macfadyen S, Hoffman AA (2010) Predicting the effects of climate change on natural enemies of agricultural pests. Biol Control 52(3):296–306
Tramberend S, Fischer G, Bruckner M, van Velthuizen H (2019) Our common cropland: quantifying global agricultural land use from a consumption perspective. Ecol Econ 157:332–341
UN DESA (2019) Growing at a slower pace, world population is expected to reach 9.7 billion in 2050 and could peak at nearly 11 billion around 2100. United Nations Department of Economic and Social Affairs (17 June 2019). Accessed 12 Feb 2020. https://www.un.org/development/desa/en/news/population/world-population-prospects-2019.html
Walther GR, Post E, Convey P, Menzel A, Parmesan C, Beebee TJ, Fromentin JM, Hoegh-Guldberg O, Bairlein F (2002) Ecological responses to recent climate change. Nature 416:389–395
Walther GR, Roques A, Hulme PE, Sykes MT, Pyšek P, Kühn I, Zobel M, Bacher S, Botta-Dukat Z, Bugmann H, Czucz B (2009) Alien species in a warmer world: risks and opportunities. Trends Ecol Evol 24:686–693
Wand SJE, Midgley GF, Jones MH, Curtis PS (1999) Responses of wild C_4 and C_3 grass (Poaceae) species to elevated atmospheric CO_2 concentration: a meta-analytic test of current theories and perceptions. Glob Chang Biol 5:723–741
Ward JK, Tissue DT, Thomas RB, Strain BR (1999) Comparative responses of model C_3 and C_4 plants to drought in low and elevated CO_2. Glob Chang Biol 5:857–867
Warren R, Price J, Graham E, Forstenhaeusler N, VanDerWal J (2018) The projected effect on insects, vertebrates, and plants of limiting global warming to 1.5 C rather than 2 C. Science 360(6390):791–795
Watt AD, Leather SR (1988) The pine beauty in Scottish lodgepole pine plantations. In: Dynamics of forest insect populations. Springer, pp 243–266
Wilson PD, Downey PO, Leishman M, Gallagher R, Hughes L, O'Donnell J (2009) Weeds in a warmer world: predicting the impact of climate change on Australia's alien plant species using MaxEnt. Plant Prot Q 24:84–87
Yan Y, Wang YC, Feng CC, Wan PHM, Chang KTT (2017) Potential distributional changes of invasive crop pest species associated with global climate change. Appl Geogr 82:83–92
Ziska LH (2000) The impact of elevated CO_2 on yield loss from a C_3 and C_4 weed in field-grown soybean. Glob Chang Biol 6(8):899–905. https://doi.org/10.1046/j.1365-2486.2000.00364.x
Ziska LH (2010) Elevated carbon dioxide alters chemical management of Canada thistle in no-till soybean. Field Crop Res 119:299–303
Ziska LH (2016) The role of climate change and increasing atmospheric carbon dioxide on weed management: herbicide efficacy. Agric Ecosyst Environ 231:304–309
Ziska LH, Bunce JA (1997) Influence of increasing carbon dioxide concentration on the photosynthetic and growth stimulation of selected C_4 crops and weeds. Photosynth Res 54:199–208
Ziska LH, Teasdale JR (2000) Sustained growth and increased tolerance to glyphosate observed in a C_3 perennial weed, quackgrass (Elytrigia repens), grown at elevated carbon dioxide. Funct Plant Biol 27(2):159–166
Ziska LH, Teasdale JR, Bunce JA (1999) Future atmospheric carbon dioxide may increase tolerance to glyphosate. Weed Sci 608–615
Ziska LH, Blumenthal DM, Franks SJ (2019) Understanding the nexus of rising CO 2, climate change, and evolution in weed biology. Invasive Plant Sci Manage 12(2):79–88

Assuring Crop Protection in the Face of Climate Change Through an Understanding of Herbicide Metabolisms and Enhanced Weed Control Strategies

Simerjeet Kaur, Khawar Jabran, Singarayer Florentine, and Bhagirath S. Chauhan

Introduction

It is essential to recognize that the current increasing levels of atmospheric carbon dioxide (CO_2), coupled with rising surface temperatures and aberrant weather situations such as excess rainfall, drought, and windstorms are becoming major concerns for those charged with the management and development of sustainable crop production. The negative impacts of these climate change dynamics on the growth of economically important crops are in addition to the increasing problems of weed infestations. In the latter case, it has been suggested that many weed species are likely to be better adapted to the effects of changes in climate than economic crop plants, due to their rich biological diversity, their inherent hardiness in the face of external climate events, and their significant persistence, even when challenged by anthropogenic control strategies.

S. Kaur
Punjab Agricultural University, Ludhiana, Punjab, India
e-mail: simer@pau.edu

K. Jabran
Department of Plant Production and Technologies, Faculty of Agricultural Sciences and Technologies, Nigde Omer Halisdemir University, Nigde, Turkey

S. Florentine
Centre for Environmental Management, Federation University Australia, Mt Helen, Ballarat, VIC, Australia
e-mail: s.florentine@federation.edu.au

B. S. Chauhan (✉)
The Centre for Crop Science, Queensland Alliance for Agriculture and Food Innovation (QAAFI), The University of Queensland, Gatton, QLD, Australia
e-mail: b.chauhan@uq.edu.au

K. Jabran et al. (eds.), *Crop Protection Under Changing Climate*,
https://doi.org/10.1007/978-3-030-46111-9_2

As a consequence, a better understanding of the climate change effects on weed control strategies is urgently required to support global agricultural industries. In particular, the changes in growth mechanisms for weeds under changing climate events needs to be appreciated, and a detailed knowledge of herbicide activity will thus be essential for effective weed control under increasing evidence of raised resistance for many weed species. Accordingly, in this chapter, we have reviewed aspects of weed growth, competition, and control under a changing climate scenario to more clearly appreciate the central issues which will concern future crop management strategies. After a brief overview of the weather effects of climate change and their consequences on agriculture, five focussed sections will offer a detailed discussion of the impacts of climate change on (i) the metabolism of various herbicides, (ii) the physiological attributes of weed species related to external weather conditions, (iii) weed growth behaviour in variable climate conditions, (iv) aspects of crop-weed competition, and (v) the ensuing implications for weed management. We have finally provided a review of anticipated actions in these areas which may enhance our understanding of the impact of climate change on the practice and future of weed management and crop production.

An Overview of Agricultural Pressures

It is well known that agricultural productivity has increased markedly since the green revolution was made possible through (i) the widespread use of high-yielding varieties, (ii) adoption of inorganic fertilizers, and (iii) the introduction of efficient irrigation facilities (Hedden 2003). Later, with the increasing infestation of crops by weed species, a fourth contribution, chemical herbicides, allowed the continuation of the expansion of the agricultural industry to meet increasing world demands.

However, the world's population is expected to exceed 9 billion by 2050 (FAOSTAT 2018) which is already putting an increased pressure on agricultural productivity and will require an increase in grain yield of about 44 million metric tons per year to meet an almost doubled food demand (Rivers et al. 2015). Exacerbating this problem is that, in the first years of the twenty-first century, sustaining the rate of agricultural growth is being challenged by many new factors such as shrinking availability of cultivable land, diverse and complex pest problems, rising production costs, diminishing economic returns, aggressive weed infestations, and the overlying effects of climate change. Clearly, pre-emptive action on all these issues needs to be taken, and we begin our discussion by looking at (i) the notion of climate change and (ii) what are its important manifestations, and then we consider the key implications for crop productivity which are within our means to regulate.

The Notion of Climate Change

Climate change is a change in the climatic variables that will persist for decades or longer periods (IPCC 2007), and one significant tracer to such change is global warming. Indeed, in this respect, Arrhenius, in the year 1896, first proposed that man-made activities (such as CO_2 emissions from coal, other fossil fuels, and land-use change) might be responsible for global warming (Weart 2003), which strongly suggests that some proactive strategies need to be developed in order to avert further damage. In the meantime, in an attempt to more clearly understand aspects related to climate change and to tackle the potential threat posed by the now evident onset of global warming, two agencies, namely, the United Nations Environment Programme and the World Meteorological Organization jointly established the Intergovernmental Panel on Climate Change (IPCC) in 1988. From their work, it has become evident, based on prehistoric climate records, that global temperatures far higher than those observed currently are indeed possible in situations of high atmospheric CO_2. We thus are concerned that this continuing and accelerating trend of warming is a stentorian warning signal for the agricultural sector. There is a widespread recognition that greenhouse gases (GHGs), such as CO_2 and methane, which are emitted due to human activities, have the potential to influence the environment in their own way, and these will have a cumulative and disastrous effect on agricultural food production and public health (Ziska et al. 2009).

Agricultural Manifestations of Climate Change

It is widely anticipated that a rise in atmospheric CO_2 levels will affect global crop production in two ways. First, there will be elevated surface temperatures due to the trapping of infrared radiation in the atmosphere, and secondly, there will be a positive increase in photosynthesis by plants due to the 'fertilization effect' (Adams et al. 1998; Valerio et al. 2011), and it is understood that these effects will interact synergistically to alter plant growth in diverse ways.

Various anthropogenic causes such as fossil fuel combustion, rice cultivation, land-use change, and the development of large herds of ruminants result into emission of GHGs. It has been reported that increases in the concentration of GHGs such as CO_2, nitrogen oxides, and methane (CH_4) closely follow the pace of growth of the global population. The atmospheric CO_2 level has risen by 24% to the current level since 1958 (IPCC 2007), but it is estimated that there has been a 35% increase in atmospheric CO_2 levels since 2000, which presents an alarming scenario (Canadell et al. 2007). It is notable that atmospheric water vapours, also a by-product of fossil fuel combustion, are the most important GHG after CO_2 and account for 60% of rising earth's temperature.

Overall, the cumulative radiative energy effect of man-made GHGs has been estimated to be 3 W/m^2 (IPCC 2007) which is equal to the effect of the earth being moved closer to the sun by a million miles. As a consequence, there will be more extreme weather conditions such as frequent heat waves and droughts, which will be generated through vagaries of the 'butterfly effect' (Mpelasoka et al. 2008). Further, projections indicate an increase in CO_2 concentration between 600 and 1000 ppm by the year 2100 (IPCC 2007) which will intensify this situation. It is estimated that the rise in global surface temperature will be between 3 and 12 °C (IPCC 2007), and this rise will vary both on a temporal and geographic scale (Cushman 1999). Global warming is predicted to occur in the winter season, and also it is anticipated that night (minimum) temperatures are going to increase proportionally more than the (maximum) day temperatures.

It is thought that moisture stress will increase due to (i) too much rainfall, as the frequency of heavy precipitation events will increase and (ii) too little rainfall, which will cause increases in aridity. Water will be more available (between 10% and 40% more) in higher latitudes, whereas sub-tropical areas will receive reduced (10–30% less) rainfall events (Gustafson 2011). The current deserts of Southern Africa, Western Australia, the Mediterranean, Eastern Brazil, Southwestern North America, and Southeast Asia will thus be expected to expand due to climate change. Such a phenomenon will necessitate the development of drought-tolerant varieties of crops which are to be grown in these areas if they are to have any chance of reaching sustainable crop production levels.

Finally, CO_2 emissions have been observed to have another serious consequence on the oceans. Dissolved CO_2 causes the formation of carbonic acid, which, although weakly dissociated, increases the acidification of sea water. This acidification, in turn, increases the solubility of a key component of the skeletons of marine organisms (calcium carbonate). Whilst the full effects of this acidification are still to be determined, it has been noted that there are reports of a recent 40% decrease in the concentration of marine phytoplankton (Boyce et al. 2010).

Climate Change and Agriculture

Whilst it is known that plants produce food from four abiotic resources, sunlight, water, nutrients and CO_2 via photosynthesis process, it is not commonly understood that it has been estimated that agricultural activities are responsible for roughly 25% of all man-made GHG emissions (Burney et al. 2010). It is the main source of N_2O (mid-season drainage) and methane emissions from rice fields (flooding), and CO_2 emissions due to general changes in land use (Gustafson 2011). The assessment of emission of methane from irrigated rice cultivation corresponds to 625 million tons of CO_2 equivalent (CO_2e). The emissions due to conventional crop chemical production (such as fertilizers, pesticides) is roughly 20 kg (CO_2e) per kg of crop chemical utilised (Wang et al. 2007).

During the time of the evolution of plants (prior to 400 million years ago), atmospheric CO_2 levels were 4–5 times greater than the present level. Concentrations have subsequently been lowered during the last 25–30 million years (Bowes 1996) with CO_2 being absorbed and used by plants during their growth cycle. However, records show that the burning of fossil fuels and land-use changes over the last two decades have increased CO_2 availability from the pre-industrial era concentration of 280 parts per million (ppm) to 379 ppm in 2005. Estimates indicate that this will increase 700 ppm by 2100 (IPCC 2007). Whilst CO_2 is beneficial for plant biology, as it is the sole supplier of carbon for photosynthesis, this sharp CO_2 enrichment might result in unforeseen physiological, morphological, and anatomical changes in the plant biome, a prediction which has its genesis in the mechanisms underpinning the relative delicate nature and balance of the natural ecosystem.

It is known that atmospheric temperature decreases by about 1 °C for every 2° increase in latitude or for each 100 m increase in altitude and that plant growth rates and rate of development processes are largely controlled by ambient air or soil temperatures. In this respect, the production of essential annual crops has been shown to be badly affected by increases in mean ambient temperatures of 2–4 °C, and thus their production is expected to decrease further towards the end of the twenty-first century in the face of further global warming. It has been estimated that the area-wise average yield of essential crops is forecasted to decrease under the slowest warming scenario by 30–46% and under the most rapid warming scenario by 63–82% by the end of the twenty-first century (Schlenker and Roberts 2009). Due to global warming, crop yields are already declining at lower latitudes and will continue to suffer more due to a consequent shortening of crop duration (Wheeler et al. 2005). In higher latitudes, it is expected that crop productivity will be enhanced until the middle of the century, whereupon increasing warming will then begin to harm the yield (IPCC 2007). It has been noted that the pattern of climate change in terms of temperature, CO_2 increase, and precipitation change is non-uniform (IPCC 2007), and thus impacts on agriculture would be expected to be highly site-specific. Of concern in this respect is that Malthus, a nineteenth-century philosopher, forewarned that unavoidable food shortages in the future would be the source of significant human catastrophe (Dyson 2001).

Climate Change and Weed Infestation

Weed infestation in cropping areas is the major biotic production problem and has been estimated to cause 34% of crop yield losses, which is a considerable effect on agricultural output (Oerke 2006). Aggressive weeds are troublesome, persistent invaders, as they are ecological opportunists and are commonly hardy and resilient plants, which show a much greater biological diversity than crop plants. They outcompete crop plants for essential resources like light, water, nutrients, and space and thus markedly reduce the crop yields. The presence of weed material in crop produce leads to a deterioration of its quality and consequently lowers the market price. It is known that yield losses due to weeds are variable depending upon the type of weeds, crop, and farming practices and that weeds also act as alternate hosts for various plant pathogens and disease organisms (Boydston et al. 2008). These factors make weed management a recognized necessity for higher crop productivity and for the maintenance of market quality (Ziska 2001).

The Impact of Climate Change on Weed Composition and Infestation Areas

Whereas it is clear that crop production is directly affected by the direct effects of climate change (temperature and water stress), there are important indirect effects related to effects of increased competition from weeds and other pest species (Korres et al. 2016). In this respect, climate change will have a marked impact on the intensity of weed pressure and will also significantly influence their range of infestation. It has also been shown that an increase in extreme events like storms, rainfall, cyclones, and associated floods may increase the rate and range of weed dispersal over considerable distances (Singh et al. 2011). In addition, due to enhanced CO_2 fertilization, generally more growth will lead to more productive and prolific plants, suggesting that insect pests, plant diseases, and weeds will all get direct and indirect benefits from global warming (Chauhan and Ramesh 2015). The sturdy and well-fed plants will attract more insect pests, and diseases will flourish under the moist and warmer seasons. Summer season crops and associated weeds will have longer growing seasons, and the survival of summer annual weeds has been predicted to increase due to global warming (Hanzlik and Gerowitt 2012). In agroecosystems, increased gene exchange may lead to an introduction of non-desired traits (such as pesticide resistance) and the development of an unwanted biotype. Further, there will be more generations of pests due to lengthened growing season, which will have a direct implication on resistance management and gene flow (Ziska et al. 2012).

Because extreme weather events like strong wind currents (Ziska et al. 2011), snowfall variability (Blumenthal et al. 2008), and high temperatures (Sasek and Strain 1990; Bunce and Ziska 2000; Ziska and Runion 2007) may aid in migration and long-distance seed dispersal through wind and water, there will be more expansion of invasive weeds to new areas (Song et al. 2009). This will result in an extra burden of weed control (Korres et al. 2016) besides other peripheral damages related to fire frequency (Blank et al. 2006) and reduction in rangeland and aquatic productivity (Ziska et al. 2011). The infestation by perennial weeds will be greater than that of annual weeds (Ziska 2014), and plant community composition will change significantly with climate change.

Adding to the complexity of this issue, it is noted that climate change can result in a positive change (such as migration of pest plants to other areas or indeed to extinction of the species) or a negative change (which will include aggressive invasion and subsequent domestication) which will have major economical and environmental consequences (Ziska and McConnell 2016). Further, due to their inherent genetic variation and plasticity, weed plants may be able to evolve 'climate change-smart' attributes (such as tolerance of extremes) more quickly than crops (Baker 1965). Climate change may also reduce or enlarge the fitness and competitive ability of the weeds and is likely to change the composition of an arable ecosystem by weed migrations. It is widely appreciated that climate, land use, and agronomic management practices all contribute to the ultimate determination of weed composition at different scales of landscape and community (Tanaka and Koike 2011). However, with changing climatic conditions being added to this mix, there may be elimination of some weed plants from the scene (local extinction) or addition of new invasive weed species (colonization). Furthermore, there may be a significant change of the 'keystone' weed species (the major or dominant weed species of an area), which will markedly transform the weed community composition (Martinez-Ghersa et al. 2000; Booth and Swanton 2002), and, by implication, also lead to a transformation of traditional management practices. It has been observed, for instance, that in one study area, weeds that were of minor importance and were opportunistic weeds such as *Stellaria media* and *Geranium* sp. and weeds belonging to Brassicaceae family have now increased their density and frequency to become a major threat (Meissle et al. 2010). Of considerable importance and concern is the finding that weed plants may adapt to a changing climate scenario by evolving new morphological, physiological, and genetic processes, resulting in evolutionary trait shifts over the population scale (Tungate et al. 2007).

Climate change, mainly through temperature increase, has induced considerable shifts in the preferred location of many species (McDonald et al. 2009) and thus caused niche and trait shifts of arable weeds (Peters et al. 2014). An outstanding example of this range shift is with *Rottboellia exaltata*, a C_4 weed, which has responded to increasing temperatures not only in terms of its increased growth, tiller production, and flowering in situ but has actively spread to previously cooler regions of the United States such as the Gulf Coast states, the lower Midwest, the South Atlantic states, and the southwest (Patterson et al. 1979).

Trait shifts are primarily seen with respect to phenology, morphology, physiology, and reproduction of a plant and are a prerequisite for niche shifts and indeed may cause range shifts by forming climatic ecotypes. Range shifts have been seen in weed plants equipped with dispersal mechanisms for migration over the landscape scale to favourable climates (Jump and Peñuelas 2005). With rise in temperature, many C_4 weeds may be encouraged to extend their geographical boundaries towards the poles or into generally cooler climates (Clements and Ditommaso 2011). For example, *Stellaria media*, *Chenopodium album*, and *Sisymbrium irio* possess strong traits for wetter and milder winters and survival, and thus populations of these weeds may increase and migrate under changing climate scenario and changed agronomic practices (Froud-Williams 1996; Fried et al. 2010; Hanzlik and Gerowitt 2012).

It is apposite to note here that the climate change-induced range shifts of weed plants cannot be clearly differentiated from man-made dispersal through transport routes or trade (Milakovic et al. 2014; Korres et al. 2015), nevertheless the weed composition of particular niche may change as a result of acclimatisation of the weed species to alternative climatic conditions in a new area, resulting in infestation shifts at the community scale (Pearman et al. 2008). It has been noted that the weed flora composition of arable ecosystems has changed due to climate change, as thermophile weeds like *Amaranthus retroflexus* and *Abutilon theophrasti*, as well as late-emerging weeds like *Chenopodium album*, *Digitaria* sp., and *Sorghum halepense*, have become more abundant in maize-based cropping systems (Peters et al. 2014).

As indicated earlier in this chapter, significant weed management strategies are necessary to maintain the potential yield of crops, and various weed control tactics such as cultural, mechanical, biological, and chemical have been adapted to maintain weed populations below critical threshold levels. Weed management options vary greatly depending upon the cropping system and the environment and are often dictated by cultural practices, education levels of farming groups, and available natural resources. We anticipate that the effects of climate change will seriously affect the management of weeds, notwithstanding the type of approach currently employed (Chauhan et al. 2014). For example, chemical control methods have been widely used in developed nations for many years, but with increasing community wealth, the use of these herbicides for weed control has also increased in developing regions in recent years (Gianessi 2013). However, the efficacy of chemical weed control does depend upon the choice of chemical being used, together with its dose, quality, target weed spectrum, time of application, and on its eventual fate in the environment and plant system. With climate change, the inevitable increased environmental variations, such as heavy rains, cyclones, and high temperatures, may markedly reduce the opportunities to chemically control weeds by changing herbicide effectiveness and persistence and increasing the potential injury to nontarget individuals (Ziska et al. 2004; Ziska and McConnell 2016).

The Metabolism of Various Herbicides

Chemical weed management is currently the most economical and widely practiced weed management practice in intensive agriculture across the world. It has multiple benefits, as it saves time, energy, and labour costs and eliminates early crop-weed competition. These chemical herbicides are classified into 26 groups based on their mode of action (Table 1, HRAC 2018 catalogue). Herbicide 'mode of action' depends upon plant functions and physiological processes required for plant growth and development and characteristically targets only a few proteins or processes like growth, amino acid synthesis, fatty acid synthesis, mitosis, and photosynthesis from the wide range of possibilities present in the intended plants. As a consequence of this mode of action strategy which targets a specific compound or process, diverse environmental conditions that affect these plant processes will naturally adversely affect the efficacy of the herbicide. However, we note here that herbicides from different groups, which are different chemical compounds, have different physicochemical properties, and thus the effect of climate change on herbicide efficacy can vary. For example, it has been suggested that any climate change effects which are

Table 1 Herbicide classification by mode of action according to the Herbicide Resistance Action Committee (HRAC)

HRAC group	Mode of action	Chemical family
A	Inhibition of acetyl CoA carboxylase (ACCase)	Aryloxyphenoxypropionates (FOPs)
		Cyclohexanediones (DIMs)
		Phenylpyrazoline
B	Inhibition of acetolactate synthase ALS (acetohydroxyacid synthase AHAS)	Imidazolinones
		Pyrimidinylthiobenzoates
		Sulfonylaminocarbonyltriazolinone
		Sulfonylureas
		Triazolopyrimidines
C1	Inhibition of photosynthesis at photosystem II	Phenyl-carbamates
		Pyridazinones
		Triazines
		Triazinones
		Triazolinone
		Uracils
C2	Inhibition of photosynthesis at photosystem II	Amides
		Ureas
C3	Inhibition of photosynthesis at photosystem II	Benzothiadiazinones
		Nitriles
		Phenyl-pyridazines
D	Photosystem-I-electron diversion	Bipyridyliums
E	Inhibition of protoporphyrinogen oxidase (PPO)	Diphenylethers
		N-phenylphthalimides
		Oxadiazoles
		Oxazolidinediones
		Phenylpyrazoles
		Pyrimidindiones
		Thiadiazoles
		Triazolinones
F1	Bleaching: Inhibition of carotenoid biosynthesis at the phytoene desaturase step (PDS)	Pyridazinones
		Pyridinecarboxamides
F2	Bleaching: Inhibition of 4-hydroxyphenyl-pyruvate-dioxygenase (4-HPPD)	Callistemones
		Isoxazoles
		Pyrazoles

(continued)

Table 1 (continued)

HRAC group	Mode of action	Chemical family
		Triketones
F3	Bleaching: Inhibition of carotenoid biosynthesis (unknown target)	Diphenylethers
		Isoxazolidinones
		Triazoles (in vivo inhibition of lycopene cyclase)
G	Inhibition of EPSP synthase	Glycines
H	Inhibition of glutamine synthetase	Phosphinic acids
	Inhibition of DHP (dihydropteroate) synthase	Carbamates
K1	Microtubule assembly inhibition	Benzamides
		Benzoic acids
		Dinitroanilines
		Phosphoramidates
		Pyridines
K2	Inhibition of mitosis/microtubule polymerization inhibitor	Carbamates
K3	Inhibition of cell division (inhibition of very long chain fatty acids)	Acetamides
		Chloroacetamides
		Oxyacetamides
		Tetrazolinones
L	Inhibition of cell wall (cellulose) synthesis	Benzamides
		Nitriles
		Triazolocarboxamides
M	Uncoupling (membrane disruption)	Dinitrophenols
N	Inhibition of lipid synthesis – not ACCase inhibition	Benzofurans
		Chloro-carbonic-acids
		Phosphorodithioates
		Thiocarbamates
O	Synthetic auxins (action like indoleacetic acid)	Benzoic acids
		Phenoxy-carboxylic-acids
		Pyridine carboxylic acids
		Quinoline carboxylic acids
P	Inhibition of auxin transport	Phthalamates
		Semicarbazones

manifested in carbohydrate production within the plant can affect herbicides which act on photosynthesis or amino acid production (Varanasi et al. 2015).

Herbicide Metabolism and Its Implications

It is known that, currently, more than 60% of the global herbicide market, in terms of economic value, is represented by herbicides from only four mechanisms of action. These are based on (i) enolpyruvate shikimate-3-phosphate, (ii) auxins, (iii) acetolactate synthase, and (iv) acetyl coenzyme-A carboxylase. Further, it has recently been commented that the herbicides belonging to these mechanisms of action have more chances of evidencing resistance development in weed species (Kraehmer et al. 2014), which could presage a significant problem for the use of chemical herbicides under climate change scenarios.

A major proportion of the reported cases of herbicide resistance in weeds to date involve herbicides belonging to Group M (glyphosate), Group A (acetyl CoA carboxylase inhibitors), and Group B (acetohydroxyacid synthase inhibitors) which are related to enhanced herbicide metabolisms (Délye 2013). It has been seen that crop plants such as wheat and maize detoxify herbicides due to enhanced activity of glutathione-s-transferase (Noctor et al. 2012) and cytochrome P450 monooxygenase enzymes (McFadden et al. 1989), and it has been predicted that similar and related detoxification mechanisms may lead to herbicide resistance in weeds (Délye et al. 2013; Han et al. 2013). It is understood, at a more detailed level, that the mechanism of development of resistance in weeds is either exclusionary resistance or target-site resistance (Dekker and Duke, 1995). In exclusionary resistance, the herbicide is not allowed to reach the site of action by physical means such as differential absorption, translocation, compartmentation or by detoxification. Herbicide uptake or absorption may be influenced by altered morphology of the plants in the form of reduced leaf area and increased waxiness of leaf surface, whereas herbicide translocation within the plant system depends upon the rate of plant growth and may be reduced under low or high temperatures due to changed water relations. The required quantity of herbicide may not reach the site of action due to compartmentation in vacuoles or enhanced metabolic detoxification by enzymes within the plant system. In each of these cases, target site resistance is achieved with either an altered site of action (e.g. an alteration in a single nucleotide in the gene encoding the protein to which the herbicide binds) or by overproduction of the site of action which leads to a considerable dilution of the effect of the herbicide. In each of these situations, plants can manage to carry on their normal metabolic activities.

Herbicide Metabolism Under Global Warming

The efficacy of chemical weed control depends on the environmental conditions before, during, and after the herbicide application. In this respect, reduced stomatal conductance of plants growing under elevated CO_2 may change the efficacy of both soil uptake and foliar uptake of herbicides. In addition, temperature variability has a direct effect on herbicide efficacy as it affects herbicide absorption, translocation, and metabolism (Robinson et al. 2015). Indeed, all plant physiological processes, including photosynthesis, respiration, and translocation, are dependent upon temperature and any change in the rate of these processes indirectly affects herbicide absorption and translocation. In this way, global warming may increase the risk of evolution of nontarget site resistance mechanisms against herbicides in the weed plants and thus decrease herbicide efficacy. Recently, decreased sensitivity to ACCase inhibitors was observed in the grass weeds such as *Avena sterilis*, *Alopecurus myosuroides*, *Lolium rigidum*, and *Setaria viridis* under elevated temperatures and, as a result, plant survival rate after herbicide application increased up to 80% (Matzrafi et al. 2016). Enhanced temperatures may also increase abiotic and enzymatic hydrolytic processes in plants, which directly affect the rate of herbicide metabolism and breakdown (Fantke et al. 2014). The levels of herbicide resistance in *Conyza* sp. to glyphosate (Ge et al. 2011), *Hordeum glaucum* to paraquat (Lasat et al. 1996), and *Lolium rigidum* to chlorsulfuron (Han et al. 2013) may all change depending upon ambient growth temperature. The level of sensitivity of the plants to the applied herbicide in a herbicide-resistant crop may also alter under varying growth temperatures. Glyphosate, a non-selective herbicide, is widely used in North America where *Roundup Ready* corn, cotton, soybean, canola, flax, and sugar beet are grown, and this compound has virtually replaced all other weed control methods in the production of these crops. In Australia, too, glyphosate is used widely to manage weeds in fallows and glyphosate-tolerant cotton. Indicating the importance of this sensitivity, it has been shown that genetically modified corn becomes more sensitive to metolachlor (Viger et al. 1991) whilst glufosinate metabolism increased in glufosinate-tolerant cotton (Mahan et al. 2006) under elevated temperatures.

Herbicide Metabolism Under Elevated CO_2

Plants adapt to elevated CO_2 through changes in plant hormonal metabolism and plant water relationships, together with alterations to carbon and nitrogen metabolism functions. It has been claimed that changes in physiology, morphology, and anatomy of plants will result in altered weed growth, crop-weed competition, and herbicide efficacy under elevated CO_2 (Ainsworth and Long 2005; Taub et al. 2008), and these changes are likely to affect the uptake, translocation, and metabolism of herbicides and thus the overall herbicide efficacy. The absorption of foliar herbicides is mainly through stomata, and decreased stomatal conductance under elevated CO_2 may result in reduced herbicide uptake of water-soluble herbicides (Ziska et al. 1999).

Atmospheric CO_2 enrichment increases the ratio of CO_2-to-O_2 at the site of CO_2 fixation and favours the photosynthetic carbon reduction cycle to produce more photosynthates and carbon intermediates (Seneweera et al. 2005). The enzymes may be present in large numbers under elevated CO_2, which is an essential requirement for other biosynthetic pathways like amino acid and fatty acid biosynthesis. This overproduction of target sites may result in the reduced efficacy of herbicides, as more copy numbers of EPSPS enzyme were observed in glyphosate-resistant plants than the susceptible plants (Gains et al. 2011). Under elevated CO_2, starch content per gram of plant tissue may be increased, but the protein content may be reduced (Bowes 1996). This suggests that there may be less demand for amino acids which may result in reduced efficacy of herbicides acting on the biosynthetic pathways of protein (Ziska et al. 1999; Varanasi et al. 2015). Further, it has been determined that increased starch content in leaves may interfere in herbicide activity (Patterson 1993).

The Physiological Attributes of Weed Species Related to External Weather Conditions

The responses of plants to rising atmospheric CO_2 levels in terms of physiological processes is dependent upon the photosynthetic pathway, and it has been shown to be highly species-specific (Ziska 2004). About 95% of the known 250,000 plant species are C_3 plants (the primary intermediate is a 3-C compound), and 3–4% are C_4 plants (the primary intermediate being a 4-C compound). The rest of known plants belong to a crassulacean acid metabolism (CAM). Of particular concern is that plants like *Parthenium hysterophorus*, which have a C_3-C_4 intermediate photosynthetic pathway, would pose severe crop-weed competition under a climate-change scenario (Chamberlain and Gittens 2004).

The oxygenation reaction catalysed by Rubisco results in 25–30% loss of fixed carbon, and this loss increases under high temperature and drought (Portis and Parry 2007). Both C_3 and C_4 plants show an increase in photosynthetic rate, but this is more pronounced in C_3 plants, since Rubisco is substrate-limited at the current atmospheric CO_2 levels. The C_3 plants have carboxylation kinetics, which is evidence of a high degree of carbon loss through photorespiration, and these respond to a greater extent relative to plants possessing a C_4 photosynthetic pathway. It is postulated that C_4 photosynthesis was evolved to adapt to high light intensities, less soil moisture, and atmospheric CO_2 (Hatch 1987), and most of the world's most aggressive weeds belong to this photosynthetic pathway (Sage et al. 2011). The C_4 photosynthetic pathway is exhibited in approximately half of the monocot species (grasses and sedges) and almost 2000 dicot species. Twelve crops out of the 15 crops which supply 90% of the world's calories are C_3 plants, and 14 weeds out of the18 world's worst weeds are C_4 plants (Patterson 1985). There are differences in reaction to raised CO_2 levels between the C_3 and C_4 species, with the intercellular CO_2 in C_3 species being 250 ppm, whilst in C_4 species it is 2000 ppm (Yoshie 1986). In addition, the carbon fixation mechanism of C_4 plants is saturated at 360 ppm (Leegood 2002) whilst that of plants belonging to the C_3 photosynthetic pathway get significantly more benefits of elevated CO_2.

It is fair to say that our understandings in this area are, as yet, far from clear. Earlier research on crop-weed competition under elevated CO_2 was based on the assumption that some of the world's the most noxious weed species possess the C_4 photosynthetic pathway and are often observed in C_3 crops (Edwards and Huber 1981). Under CO_2 enrichment, in one study, it was found that a significant increase in photosynthesis of C_3 plants was seen along with no change in photosynthesis of C_4 plants (Ziska et al. 1999), but in a separate investigation, a similar increase in photosynthesis of both C_3 and C_4 plants was observed (Ziska and Bunce 1997). Further, photosynthetic stimulation was observed to be doubled in C_4 weed plants compared to C_3 crop plants at the higher CO_2 environment (Ziska and Bunce 1997). As a consequence, the generalized assumption that a C_3 plant will get more benefit from CO_2-enriched environment may not be accurate. Barnaby and Ziska (2012) have reviewed the variable relative responses of C_3 and C_4 species in term of photosynthetic stimulation, but the physiological basis for the variability is unclear, and the stimulation may continue for weeks or months (Ziska and Bunce 1997).

There are other examples of the complexity of this area. Whilst it should be appreciated that CO_2 enrichment will influence vegetation dynamics irrespective of whether it is a crop or weed (Kimball et al. 2002), the theoretical increase in net photosynthesis was estimated to be 38% in C_3 plants (Long et al. 2004). However, in free-air CO_2 enrichment experiments (FACE), only a 14% increase in photosynthesis was observed under a CO_2-enriched environment (Long et al. 2006). There was no effect of CO_2 enrichment on the photosynthetic rate of C_4 plants like *Cyperus rotundus* and *C. esculentus*, but it did increase their instantaneous water use efficiency (Rogers et al. 2008). It has been observed that there are definite changes in competitiveness between C_3 weed, common cocklebur (*Xanthium strumarium* L.), and C_4 crop (*Sorghum bicolor* L.) under high CO_2 as the reduction in single leaf photosynthesis was less in C_3 weed than C_4 crop plant under CO_2 enrichment when grown in a mixture (Ziska 2001). It is also known that elevated CO_2 has enhanced maximum photosynthetic rate, net photosynthetic efficiency, and water use efficiency of *P. hysterophorus* (Pandey et al. 2003), and it has been stated that the plant growth reduction of crops and weeds is mainly due to a reduction in leaf area duration and less due to net assimilation ratio under CO_2 enrichment when grown in competition (Patterson et al. 1984).

Stomatal conductance and transpiration rate decreased with an increase in CO_2 from 375 ppm to 675 ppm (Patterson et al. 1982), and leaf osmotic potential decreased at a faster rate under elevated CO_2 levels than for the ambient CO_2 counterparts (Sionit et al. 1980). Water use efficiency increased with reduced transpiration, and this was thought to be due to stomatal closure under CO_2 enrichment. The net assimilation rate was also seen to increase, and this is possibly due to the diffusion pathway of water vapour and CO_2 being very similar. Whilst with CO_2 enrichment, stomatal number, and conductance decreased in both C_3 and C_4 plants (Ainsworth and Long 2005), the effect of CO_2 enrichment on photosynthesis is variable in C_4 plants (Ziska and Bunce 1997). The C_3 plants obtain benefits both from improved photosynthesis function and water use efficiency, whilst C_4 plants get advantage only from improved water use efficiency under elevated CO_2. Improved water use efficiency may result in the evolution of more drought-tolerant plants (Valerio et al. 2011).

Nutrient Variations

In the anticipated future carbon-rich environment, the relative availability of nitrogen and phosphorus will become limiting factors to plant growth (Hall and Allen Jr 1993). It has already been observed that whilst the carbohydrate content per gram of plant tissue increased under CO_2 elevation, the content of protein was reduced (Kimball et al. 2002). This suggests that the nutritive value of plants, in particular essential crops, may decrease.

Of concern is that whilst transpiration rate and stomatal conductance were not affected by nutrient levels, the leaf chlorophyll content was significantly decreased under low nutrient supply (Patterson et al. 1982). This suggests that elevated CO_2 may intimately affect nutrient cycling and microbial activity (Patra et al. 2012), and it has been claimed that the effect of variation in the nutrient level on plant development was higher and more consistent than variation in CO_2 levels (Patterson et al. 1982). It has also been noted that reductions in stomatal conductance and transpiration have evidenced a negative effect on water and nutrient uptake, and this could mean that C_4 plants will respond more positively under a reduced N availability scenario since they have an inherently higher nutrient use efficiency than C_3 plants (Bleier and Jackson 2007). In addition, it has been shown that the photosynthetic nitrogen use efficiency is greater for C_4 plants than C_3 plants under higher temperatures (Long 1999).

Effects of Warming on Physiological Parameters

Climatic change is associated with a higher frequency of extreme weather events, and it is generally agreed that this will have a drastic impact on ecosystem productivity and biodiversity. Plant growth rate depends upon the ambient temperature, which is the primary factor determining early plant growth (Storkey and Cussans 2000). As mentioned earlier, a characteristic of weed species is that they adapt well to all climatic stresses due to their inherent hardiness, and this indicates that they can develop various strategies including accumulation of heat shock proteins, osmoprotectants, antioxidants, and alterations at the molecular level to offset stress-induced biochemical and physiological changes (Hasanuzzaman et al. 2013). It has been noted that the complex interaction between elevated CO_2 and temperature may limit the antitranspirant effects of CO_2, and rising temperatures may override the stimulating effects of elevated CO_2 on plant growth in C_3 plants (Batts et al. 1997). In comparison, the C_4 photosynthetic pathway will provide plants with a major advantage under hot, arid, and high light conditions, but it is less efficient at temperatures below 25°–28 °C under light-limiting conditions (Lara and Andreo 2011). Recent work has suggested that ambient warming may lead to increased solution-phase nitrogen and N mineralization, and this will have a marked effect on plant growth (Patra et al. 2012). In addition, elevated ambient temperatures will result in higher transpiration rates, less biomass production, and shortened reproductive development leading to lower seed production (Adams et al. 1998).

These findings suggest that the growth of weed plants may significantly increase under higher ambient temperatures, which indicates that infestations of weeds can become seriously problematic for agricultural systems in a warmer environment. In this respect, the plant biomass of *Imperata cylindrica* increased when plants were grown at 29/23 °C than at 23/17 °C, and this change was suggested to be due to the significant effects of heightened warming on the leaf area (Patterson et al. 1980). In a similar experiment, it was shown that two weedy grass species, *Panicum texanum* and *Panicum miliaceum*, which are used as food grain and forage, responded well to warming, with greater tiller numbers and biomass being produced when plants were grown at 30/24 °C than at 24/18 °C (Patterson et al. 1986). As a consequence, it is predicted that these weeds are likely to invade the new areas at higher latitudes and also to expand their geographical range with the expected rise in temperatures.

Effects of Water Stress

More frequent and severe drought stress events are predicted as a result of increasing climate change. Of relevance here is that stomatal conductance in plants has been shown to be reduced by 20% when grown under a CO_2-enriched environment (Long et al. 2006). Water deficit may also result in stomatal closure which will result in a reduced CO_2/O_2 ratio in leaves and photosynthesis (Griffin et al. 2004). Because weeds inherently possess excellent drought tolerance, they could clearly pose an increased risk for crop production under water-scarce conditions. An example is with *Parthenium* plants, which have a demonstrated higher water use efficiency, suggesting that this species will become a noxious weed under a climate-change scenario (Moore et al. 1987; Pandey et al. 2003; Shabbir et al. 2014). Also, it has been noted that there was a greater relative increase in stomatal resistance in C_4 plants than C_3 plants under elevated CO_2 levels (Akita and Moss 1972), and a well-watered C_4 weed (*A. retroflexus*) exhibited increased photosynthesis under elevated CO_2 levels. In general, C_4 plants have an advantage over C_3 plants in response to elevated CO_2 and more frequent and severe droughts (Ward et al. 1999; Valerio et al. 2011). In a separate investigation, a well-watered C_3 weed

(*Abutilon theophrasti*) exhibited a significant increase in leaf area under elevated CO_2 levels, but, under drought conditions, it appears that C_3 plants had more leaf area when grown at pre-industrial CO_2 levels than C_3 plants grown at elevated CO_2 levels (Ward et al. 1999; O'Donnell and Adkins 2001).

Weed Growth Behaviour Under Variable Climate Conditions

It is generally agreed that environmental stresses affect a plant's ability to achieve its potential growth and yield (Patterson 1995a), and it is common that, in a normal agroecosystem, each crop will need to compete with 8–10 troublesome weeds. This is a stressful situation because weeds have high biological flexibility and ecological adaptability, and the experienced increase in CO_2 will have a direct and important influence on plant growth. Since C_3 systems are deficient in carbon, CO_2 enrichment will cause a 'fertilization effect' and markedly increase plant growth for both C_3 crops and weeds (Patterson and Flint 1980). It is known that whilst many weed species possess the C_4 photosynthetic pathway, most crops are based on the C_3 pathway. This perspective meant that earlier studies in weed research were focused mainly on the outcomes of C_3 crop-C_4 weed interactions. Studies indicated that the growth of C_4 plants can be either stimulated (Bazzaz et al. 1989; Manea et al. 2011) or remain unaffected (Erickson et al. 2007) or indeed even decline (Belote et al. 2003). Nevertheless, generally under elevated CO_2 levels, the relative increase in plant biomass was more in C_3 weeds than for C_4 weeds (Bazzaz and Carlson 1984; Ziska 2000). However, later work, looking at the interplay between drought and high temperatures, has indicated that these conditions favour the competitive ability of C_4 weeds over C_3 weeds (Fuhrer 2003), which again highlights the innate complexity of this area.

Further work has indicated that there was 55% and 62% increase in plant biomass and leaf area, respectively, in red/weedy rice (*Oryza sativa*) biotypes when subjected to an increase in CO_2 level to about 550 ppm (Ziska and McClung 2008). The effects of elevated CO_2 may be due to an altered phenology of plant development phases such as is shown by the accelerated maturity rate in *Avena fatua* (O'Donnell and Adkins 2001). Research has indicated that there is greater genetic diversity and physiological plasticity amongst weedy and wild lines relative to domesticated crops (Treharne 1989), and this implies that any change in abiotic resources should favour weeds over crop species (Vengris et al. 1955; Carlson and Hill 1985). Further, genetic diversity exists amongst biotypes of same weeds (near-isogenic lines), and in this circumstance a variable plant growth response (increase in plant dry weight from 32% to 87%) was observed in *A. fatua* to elevated CO_2 level (O'Donnell and Adkins 2001). It is suspected that this problem of weedy rice may increase in Asian countries under a climate-change scenario (Chauhan et al. 2014), and this suggests that more options for controlling this weed may be required in future.

Further complications in understanding this area is illustrated by the observation that the response to CO_2 fertilization decreases with the plant age in annual plant species (Stitt 1991; Ziska and Bunce 1995), and, additionally, more vegetative growth is observed relative to reproductive growth. Also, the reproductive capacity in weeds may increase in response to CO_2 fertilization (Patterson et al. 1999; Ziska 2004, Ziska and Runion 2007), and elevated CO_2 may cause more flower production [e.g. in *Amaranthus viridis* (Naidu and Paroha 2008)] and pollen production [e.g. in *Ambrosia artemisiifolia* (Wayne et al. 2002)]. It has been shown to be a significant increase in vegetative biomass of plants grown in a CO_2-enriched environment, but this increase may not be reflected in a higher seed yield, and, indeed, inconsistent effects on seed yield have been observed (Ziska and Goins 2006).

It is becoming clearer therefore that perennial weeds might be a greater concern for crop production under elevated CO_2 levels than annual weeds. Moreover, perennial weeds have been observed to flourish more frequently under continuous zero-till fields than annual weeds (Gibson et al. 2005). Evidence to this effect comes from the observed stimulation of photosynthesis and growth under long-term exposure of elevated CO_2 levels which was sustained in *Elytrigia repens,* a perennial C_3 weed (Ziska and Teasdale 2000). In this work, no photosynthetic acclimation was observed, unlike with experiments with other herbaceous agronomic perennials such as alfalfa (*Medicago sativa*) and orchard grass (*Dactylis glomerata*) (Bunce 1995). The increase in weed biomass of *Cyperus rotundus* was 33–40% whilst that of *C. esculentus* was 10–15% when exposed to elevated CO_2 (Rogers et al. 2008). With CO_2 levels elevated from 380 to 670 ppm, a significant increase in aboveground biomass of *Cirsium arvense* was observed, and the establishment of *C. arvense* was found to be more as a function of CO_2 levels as 30% less number of cuttings were required to reestablish the weed over time (Ziska 2010). The root to shoot ratio was improved with increasing CO_2 in C_3 plants, but no such effect was observed in C_4 plants (Patterson and Flint 1980). There was more partitioning of photosynthates to roots than shoots under elevated CO_2 in perennial weeds like *Cyperus* spp. (Rogers et al. 2008) and *Cirsium arvense* (Ziska et al. 2004), and an increased root to shoot ratio was consequently observed.

The growth of invasive weed species is dependent upon climate change and is consequently found to be increasing (Ziska et al. 2011). The variable rate of stimulation of plant growth of six invasive weed species of the United States, *Cirsium arvense*, *Convolvulus arvensis*, *Euphorbia esula*, *Sonchus arvensis*, *Centaurea maculosa*, and *Centaurea solstitialis* (all C_3 plants), was observed when grown under a CO_2-enriched environment with the largest response observed in *Cirsium arvense* (Ziska 2003a). The invasive weed species like *Bromus tectorum* respond strongly to CO_2 enrichment and is responsible for firebreaks (Ziska et al. 2005).

Although the role of weeds in the economically important agricultural production system is an essential topic for weed scientists to study, it is also relevant, and indeed crucial, for studies to be intensified in the area of growth behaviour of those weeds which cause human health problems (Ziska and Caulfield 2000; Ziska 2003b). In this respect, the pollen production of *Ambrosia artemisiifolia* was enhanced by 131% and 320% when plants were exposed to 370 ppm and 600 ppm, respectively, relative to 280 ppm of CO_2. This clearly indicates that continuous increase in atmospheric CO_2 will directly affect public health (Ziska and Caulfield 2000), in this case in the exacerbation of asthmatic conditions. *P. hysterophorus* plants, which are recognized as being injurious to human health, were 52% taller and had 55% more weed biomass when observed under elevated CO_2 (Shabbir et al. 2014). Future elevated CO_2 levels are likely to expand the area of land under *Toxicodendron radicans*, and the plant growth, rate of rhizome elongation, and production of urushiol (oil) content (which is responsible for dermatitis) will be stimulated (Ziska et al. 2007).

Effects of Warming on Weed Proliferation and Growth

Along with CO_2 increase, the temperature of the earth is seen to be increasing, and this temperature change is anticipated to significantly affect the geographical distribution of weeds (Patterson et al. 1999). The ambient temperature plays a key role in deciding the new environments for colonization of a plant species (Woodward and Williams 1987), and it has been noted that increasing temperatures affect plant phenology more than the elevation of CO_2 levels (Lee 2011). It has been established that, under increasing ambient temperatures, weed plants move forwards to higher latitudes (Patterson 1995a) and upwards to higher altitudes (Parmesan 1996). In this respect, the minimum temperature in winter may be a significant driver of weed shifts (Ziska et al. 2011), and, in addition, warming could make weed plants more competitive than crop plants (Rosenzweig and Hillel 1998). Higher temperatures strongly affect the biomass production at the reproductive stage compared to the vegetative stage (Lee 2011), but they have a negative effect on pollen viability, pollen tube germination, fertilization, and seed formation. The growth of *P. hysterophorus* in term of crop biomass, crop growth rate, relative growth rate, net assimilation rate, leaf area duration, leaf area index, and numbers of flowers and seeds was improved at elevated temperature. During this work, the maximum photosynthetic response was observed at 25–35 °C, whilst the net photosynthesis was appreciably reduced at the lower temperature of 7 °C (Pandey et al. 2003).

Elevated temperatures affect the growth of C_3 plants to a greater extent than for C_4 plants, but this adverse effect for C_3 plants can be attenuated by elevated CO_2 (Lee 2011). The simultaneous effects of warming and elevated CO_2 levels will more positively favour C_4 plants relative to C_3 plants, a phenomenon illustrated by a study that showed that at mild temperatures, elevated CO_2 levels favoured rice, a C_3 crop, whilst, at a higher temperature and CO_2 levels, *Echinochloa glabrescens*, a C_4 weed, was favoured (Alberto et al. 1996). Rising temperatures may result in altered phenology of plant growth phases like leaf development, flowering, fruiting, and vernalization period (Baldocchi and Wong 2008) and thus was also shown by seedling emergence being advanced by 26 days in C_3 (*Chenopodium album*) and by 35 days in C_4 (*Setaria viridis*) weeds (Lee 2011) under warmer conditions. The warmer temperature may accelerate the growth phenology and hasten the plant's maturity

(Singh et al. 2011). It is clear that temperature increase favours the growth of C_4 weed species, as there was 68% and 88% increase in leaf area and biomass of *Rottboellia cochinchinensis* with an increase in temperature by 3 °C (Ziska and Bunce 2007). It has been further shown, in Southern India, that there is more growth partitioned to the underground plant parts of *Prosopis juliflora* with an increase in temperature (Walter and Armstrong 2014; Kathiresan 2017), whilst the abundance and growth of the parasitic weed, *Striga lutea*, may increase under the elevated temperature conditions in the United States (Patterson et al. 1982).

The Effects of Water Stress on Weed Distribution

Water deficit has been demonstrated to lower the amount of dry matter accumulation by crop plants and to shorten the life style of plants, particularly those belonging to the C_3 photosynthetic pathway (Blum 1996). In separate studies, deep-rooted plants, for example, perennial weeds, were shown to be at a competitive advantage relative to shallow feeder plants under water stress (Stratonovitch et al. 2012), whilst in areas receiving higher precipitation, anoxic conditions will be created and shallow-rooted plants will be favoured (Patterson 1995a). Under conditions of soil moisture stress, the relative increase in plant dry weight and root dry weight was more than elevated CO_2 level of 480 ppm than at the lower 357 ppm CO_2 (O'Donnell and Adkins 2001). An increase in water use efficiency was also observed at elevated CO_2 levels, but it was found that these increases did not mean that plants will necessarily use less water over the whole growing season (Morrison and Gifford 1984). Further, whilst the negative effect of water stress on plant growth was observed at both 20/16 °C and 23/19 °C, the amount of plant height was greater at the higher temperature. As a result of such observations, it is expected that weeds like wild oats will continue to grow and complete their life cycle under predicted harsher and drier extremes in future, although there will be a concomitant reductions in plant growth and fecundity (O'Donnell and Adkins 2001).

From a plant ecology point of view, it has been established that the typical succession trajectory is always from ruderals to competitive plants, then finally to stress-tolerant species. It is also known that under any type of challenging habitat, weed plants can dominate under high-stress intensity due to their inherent hardiness (Korres 2005). In such conditions, CO_2 enrichment may reduce the negative effects of water stress and increase the growth of both C_3 and C_4 plants by improving their water economy, but it has been noted that growth stimulation was more evident in C_3 plants (Patterson 1986).

Aspects of Crop-Weed Competition Under Stressful Climatic Change Situations

Climate change may have a differential effect on the growth and fitness of weeds and crops and will thus change the nature of crop-weed competitive interactions (Patterson 1995a). The particular responses to climate change depend upon the characteristics of the crop-weed association and the competitive abilities of the plant in mixed cropping situations. Naturally, any factor which aggravates the stress on crop plants may make them more vulnerable to intense weed competition, whilst the response of weeds is typically more resilient. However, climate change is also associated with atmospheric CO_2 increase, which has the previously mentioned fertilization effect on crops belonging to similar photosynthetic pathways and an improved weed to crop biomass has been observed in some instances (Ziska and Runion 2007; Ziska et al. 2011). This observation was consistent with the commonly accepted axiom that C_4 weeds will show minimal response to CO_2 enrichment, a finding that was based on earlier C_3 crop-C_4weed competition investigations. Overall, these published reports confirmed that C_3 crops became more competitive relative to C_4 weeds in response to CO_2 enrichment (Patterson and Flint 1980; Patterson et al. 1984; Patterson 1986; Alberto et al. 1996; Rosenzweig and Hillel 1998; Ziska 2000; Bunce and Ziska 2000).

However, notwithstanding the body of evidence that suggests that weed competition in a C_3 crop-C_4 weed scenario will decrease under elevated CO_2 concentration, this should not be viewed as a universal crop/weed axiom, since the situation is very complex. It appears that plant species belonging to similar photosynthetic pathways (Patterson 1995b; Wand et al. 1999; Ziska and Bunce 1997; Miri et al. 2012) or families (Wand et al. 1999; Dikšaityte et al. 2014) may respond differently to abiotic resources such as CO_2 enrichment, soil water content, and N supply (Kimball et al. 2002; Zhu et al. 2008). Legumes, for example, can respond more positively to elevated CO_2 concentration than other plant species due to the presence of large carbon sink (N fixing bacteria) in their roots (Rogers et al. 2009; Leakey et al. 2009). Some studies, which have examined the crop-weed competition between C_4 crops and C_3 weeds, have actually observed more weed competition under elevated CO_2 levels (Ziska 2001, 2003c). For example, the vegetative biomass of *Xanthium strumarium*, a C_3 weed, was increased relative to that of sorghum, a C_4 plant, at elevated CO_2 (Ziska 2001). The grain yield and biomass of sorghum was significantly reduced when grown in C_4 crop-C_3 weed (*Sorghum bicolor-Abutilon theophrasti*) or a C_4 crop-C_4 weed scenario (*Sorghum bicolor-Amaranthus retroflexus*) under CO_2-enriched environment, indicating that there will be potentially more yield loss of a C_4 crop in weedy environments as atmospheric CO_2 increases (Ziska et al. 2003c). Finally, it has been seen that elevated atmospheric CO_2 levels may enhance tolerance of rice against parasitic weeds, namely, *Rhamphicarpa fistulosa*, *Striga hermonthica*, *S. aspera*, and *S. asiatica*, in Sub-Saharan Africa (Rodenburg et al. 2011).

Crop/Weed Competition Under Global Warming Conditions

It has already been observed that increased surface temperatures will accompany elevated atmospheric CO_2 concentrations (IPCC 2007). Whilst this elevated CO_2 has been seen to increase the crop biomass of C_3 crops relative to C_4 weeds in different studies, it is suggested that abiotic stresses such as increased temperature could negate this advantageous effect in C_3-C_4 competitions (Knapp et al. 1993). For example, the grain yield of rice was reduced at 37/29 °C compared to that at 27/21 °C irrespective of CO_2 concentration, and the competitive ability of *E. glabrescens*, a C_4 rice weed, was enhanced (Alberto et al. 1996). It has been seen that simultaneous increases in both factors resulted in increased pollen sterility in rice, and the competitiveness of rice relative to C_4 weeds was decreased under elevated CO_2 levels and increased warming (Alberto et al. 1996). In tomatoes, warming (from 21/12 to 26/18 °C) resulted in increase in crop losses from competition by weeds from 33% to 55% and 32% to 61% under ambient and elevated CO_2, respectively, this loss being irrespective of photosynthetic pathway of the weed. The crop duration was also shortened (Valerio et al. 2013). Cotton was a strong competitor to *Anoda cristata* at a day/night temperature of 32/23 °C, whereas *Anoda cristata* was a superior competitor at 26/17 °C (Flint et al. 1983). The biomass of cotton increased by 38% and 61% when grown at 26/17 °C and 32/23 °C, respectively, and elevated CO_2 concentration of 700 ppm as compared to 26% and 41% increase in associated *Anoda cristata* weed (Patterson et al. 1988), indicating that cotton will be at a competitive advantage in CO_2 rich environments even under suboptimal temperatures.

Crop/Weed Competition Under Water Stress Conditions

Generally, crop-weed competition for water increases under low soil moisture conditions, and thus weed management remains important under water-scarce conditions, even though under limited water conditions, the actual impact of weed competition on crop yield appears to be lowered (Mortensen and Coble 1989). Nevertheless, the critical period of crop-weed competition has been shown to be more under wet conditions rather than dry periods (Coble et al. 1981). Wiese and Vandiver (1970) reported that plant species grown in irrigated conditions were most competitive under moist conditions, whilst those grown in drier conditions were least competitive at high moisture levels. C_3 species were least productive under dry conditions, whilst the productivity of C_4 species declined at high water availability (Ozturk et al. 1981). Elevated CO_2 under well-watered conditions increased plant height, leaf area, and biomass of tomato, a C_3 crop, more than *A. retroflexus*, a C_4 weed, both in monoculture and competition situations, but the reverse was true under water-limited conditions (Valerio et al. 2011). There was non-significant effect of CO_2 and water stress on C_4 weeds under a monoculture scenario, but when grown in competition, C_4 weeds benefitted from elevated CO_2 due to high leaf water potential compared to C_3 crops under drought conditions (Knapp et al. 1993; Valerio et al. 2011).

Effects of Variable Nitrogen Supply

Weed growth is generally stimulated under application of nitrogenous fertilizers (Vengris et al. 1955), and numerous studies have confirmed that whilst there is a stimulatory effect of CO_2 on rice (a C_3 plant) growth and yield in monoculture, crop response to elevated CO_2 depends upon N supply (Anten et al. 2003; Zhu et al. 2008). However, there is possibility that growth of C_4 species might be less dependent upon nitrogen supply due to their inherent greater nitrogen use efficiency (Sage and Pearcy 1987; Zhu et al. 2008). In the future, it is anticipated that there will be elevated concentration of nitrogen in soils (due to change in carbon and nitrogen mineralization pathways under elevated temperatures) and CO_2 in the atmosphere and, in this respect, elevated levels of CO_2 from 400 ppm to 700–1400 ppm increased the photosynthetic parameters and growth of C_3 pea crops (*Pisum sativum* L.), whilst stimulatory effects of increased N supply was observed in *Melilotus alba* Medik of the same family (Dikšaityte et al. 2014). In competition studies of C_3 crop (rice) and C_4 weed (*Echinochloa crus-galli*), the proportion of crop biomass increased with elevated CO_2 under adequate N supply relative to weed, but under low nitrogen supply, it was observed that elevated CO_2 significantly reduced the leaf area and root biomass of rice crops in comparison to the C_4 weed (Zhu et al. 2008). The role of nitrogen in moderating the stimulatory effect of elevated CO_2 has not been fully understood, but it has been proposed that it might be required for sink production (output such as seed, rhizome) (Ziska et al. 1996), since there is down-regulation of photosynthesis even under elevated CO_2 in the absence or low supply of nitrogen due to lack of sinks in C_3 crops (Anten et al. 2003). These types of studies confirm the fact that rising atmospheric CO_2 could still aggravate the crop-weed competition under limited nitrogen supply.

Effects of CO_2 Enrichment

In most common agroecosystems, economic crops compete with approximately 8–10 dominant weeds (Bridges 1992), and these weeds will belong to C_3, C_4, or CAM pathways. Most studies on crop-weed competition in CO_2 elevated conditions have focused on the alterations to the photosynthetic pathway systems of the crops with respect to weeds. In a typical study, the crop yield loss observed for soybean production increased from 28% to 39% when the crop was grown in competition, under elevated CO_2 levels, with *Chenopodium album*, which is a C_3 weed. By comparison, the reduction in yield declined from 45% to 30% when the crop was grown in competition with *A. retroflexus*, a C_4 weed (Ziska 2000). In a separate study, soybean, a C_3 crop, responded better to CO_2 elevation than C_3 weeds under monoculture conditions, but the relative yield of soybean decreased when in competition with *C. album*, a C_3 weed (Miri et al. 2012). An earlier study had indicated that soybean growth increased in competition with *Sorghum halepense*, which is a C_4 weed (Patterson et al. 1984). Wheat, a C_3 crop, responded more to CO_2 enrichment in competition with some C_3 weeds, but the response was variable depending upon the nature and type of associated C_3 weed species (Naidu and Varshney 2010). The plant relative yield of millet, a C_4 crop, increased both in monoculture and in competition with *Amaranthus* sp., a C_4 weed, with increased CO_2 concentration (Miri et al. 2012).

The relative plant performance in term of shoot biomass and leaf area of C_3 weed improved significantly when grown in competitive mixture with a C_4 crop under elevated CO_2 concentrations (Ziska 2001). Also, a significant reduction in seed yield and crop biomass of a soybean crop was observed in presence of *Cirsium arvense*, and this reduction was even higher under elevated CO_2 levels (Ziska 2010), whilst elevated CO_2 levels from 400 to 800 ppm in tomato crops exacerbated the weed competition from both C_3 and C_4 species (Valerio et al. 2013).

We note here that it is important to appreciate that the crop and weed may have similar genetic, developmental, or phenotypic characteristics such as in the case of oats (*Avena sativa* L.) and wild oats (*A. fatua*) or taxonomical and biochemical similarities like rice (*Oryza sativa* L.) and weedy/red rice (*Oryza* spp.). There was a significant decrease in relative plant yield of cultivated rice seed yield in the presence of weedy rice with an increase in CO_2 level from 300 to 500 ppm (Ziska et al. 2010), and it was commented that the higher growth yield was observed in response to increasing CO_2 level in the vegetative rather than the reproductive stage in crop lines (Ainsworth et al. 2002).

The Ensuing Implications for Weed Management Under Climatic Changes

It has been clearly stated that weed management strategies must adapt to both weed biology and ecology and impacts of climate variability (Singh et al. 2011). In these considerations, it must be appreciated that the effectiveness of chemical weed control depends upon the amount of herbicide applied, the growth stage of weeds, and the nature of the surrounding environmental factors before, during, and after the herbicide spray (Ziska and Dukes 2011). Climatic factors such as temperature, rainfall, wind, and relative humidity affect the herbicide efficacy, particularly in the case of foliar uptake herbicides (Kudsk and Kristensen 1992). It has also been claimed that unfavourable weather may either reduce herbicide effectiveness or increase the risk of crop phytotoxicity due to herbicides (Patterson et al. 1999).

Crops and weeds respond differently to climate change pressures, and under these changing conditions, it must be recognized that competing weed composition may also change. The variable response of plants to climate changes have been seen to be due to morphological, anatomical, and physiological changes under climate change, depending upon the photosynthetic pathway, available soil nutrients, and soil water profile. As a consequence, the fecundity and evolutionary rate of feral plant species will be altered, and the resultant management of wild/weedy plants will thus be difficult under climate change (Shivrain et al. 2007, 2009). It has been seen that there is more genetic diversity and phenotypic plasticity in weedy rice biotypes, and these have been shown to be able to exploit available abiotic resources far more than the cultivated rice biotypes (Ziska and McClung 2008). Of particular concern is that herbicide resistance problems in weeds are increasing at an alarming rate and many resistant cases related to common herbicides are being reported from around the world (Heap 2018). In this respect, it is thought that the sole reliance on weed management with herbicides, usually with repeated applications of single herbicide in a single season and a lack of integration with other biological and cultural methods, has contributed to the accelerated evolution of herbicide resistance.

Problems Arising from CO_2 Fertilization

The morphological, physiological, and anatomical changes in the plant system due to CO_2 enrichment could result in altered absorption, translocation, or metabolism of herbicides and overall effectiveness of herbicides (Ziska and Teasdale 2000; Ziska 2016). Water-soluble herbicides are absorbed mainly through stomata and cuticles, and CO_2 enrichment results in stomatal closure due to increased guard cell Ca^{2+} concentration (Webb et al. 1996) and reduced stomatal conductance, which may reduce the absorption of water-soluble herbicides. The lowered stomatal conductance leads to reduced transpiration rates (Ainsworth and Long 2005), thus the apoplastic movement of herbicide may be negatively affected in plants grown under elevated CO_2. Plants grown at elevated CO_2 have more leaf cuticular thickness and leaf pubescence (Ainsworth and Long 2005), and this has been shown to retard herbicide absorption. There is also a greater development of leaf area, a faster rate of leaf development, and more biomass production in plants grown under elevated CO_2 (Conroy et al. 1994), which will lead to more absorption of foliar herbicides. More photosynthesis will lead to more translocation of assimilates from source to sink tissues (Fernando et al. 2016) which, in turn, will enhance the symplastic movement of herbicides in the plant system. CO_2 enrichment increases the starch content in leaves (Robinson et al. 2012) and decreases N concentration of leaves (Högy et al. 2009; Robinson et al. 2012) and in grains (Högy et al. 2009; Fernando et al. 2015). The importance of this finding is that more starch in leaf tissues may hamper the diffusion of herbicides to a specific site of action (Ziska and Teasdale 2000), thus affecting herbicide translocation. There may be less demand for amino acids due to low protein in plant tissues which will reduce the efficacy of herbicide acting on amino acid biosynthetic pathways. This may have serious implications beyond weed control as less gluten protein in wheat grains will result in lowered bread quality and poor nutritional quality (Fernando et al. 2015).

In higher CO_2 environments, there will be a change in plant phenological development, such as a reduced length of the seedling stage (early vegetative growth), synchronicity in flowering, and spatial proximity. The weed plants cross the herbicide-sensitive stage very rapidly due to rapid vegetative development (Ziska et al. 1999), which also may decrease herbicide efficacy. The changed flowering pattern under changing biotic and abiotic factors can lead to gene transfer between crop and related wild/weedy species (Ellstrand et al. 1999). The weedy rice can be controlled with transplanting nursery in flooding fields or by cultivating non-transgenic imidazolinone-resistant rice cultivars (Gealy et al. 2003), but CO_2 enrichment will affect the management of this weed due to changed plant phenology. A 0.22–0.71% increased flow of genes was observed from wild, weedy rice to herbicide-tolerant rice when atmospheric CO_2 level was increased from 300 to 600 μmol mol^{-1}, and this increased outcrossing in response to rising CO_2 was due to advancement in flowering time for the wild rice by 8 days (Ziska et al. 2012).

The efficacy of glyphosate has been reported to decrease under elevated CO_2 levels for C_3 weeds (Ziska et al. 1999, 2004; Ziska and Teasdale 2000; Archambault 2007; Ziska 2010) and C_4 grasses (Manea et al. 2011), but growth and management of *A. retroflexus*, a C_4 broadleaf weed, was not affected by CO_2 enrichment (Ziska et al. 1999). The reasons for decreased efficacy of glyphosate appear to be numerous, including the decreased length of the herbicide-sensitive weed stage (Ziska et al. 1999), more aboveground weed biomass resulting in a dilution effect (Ziska 2010; Manea et al. 2011), an increased root-shoot ratio (Ziska et al. 2004; Rogers et al. 2008), more regeneration capacity of perennial weeds from vegetative means under elevated CO_2 levels (Donald 1990; Ziska 2010), or reduced need of aromatic amino acids (Bowes 1996).

It has been found that perennial weeds are more effectively controlled with mechanical methods like tillage and removal using cutting and hoeing. However, elevated CO_2 will increase root to shoot ratio, and this will facilitate more regeneration of new plants from tubers, rhizomes, suckers, runners, and plant fragments. This increased regenerative capacity and more belowground plant biomass under elevated CO_2 may then lead to poor control of these perennial weeds with mechanical methods, and the physical process of tillage may help in dispersal of these weeds over space and time instead of helping to control them (Ziska and Bunce 2007; Rodenburg et al. 2011).

It is suspected that biological weed management will also be affected by climate change, and the actual efficacy of biocontrol agents may also be disrupted by altered weed biology and ecology under elevated CO_2 levels. Insects, when used as biological control agents, may also fail to control the weeds due to more vegetative growth in response to elevated CO_2 levels. However, it has been noted that the stem galling moth (*Epiblema strenuana*) reduced the biomass production of *P. hysterophorus* both at ambient (by 36%) and elevated CO_2 (by 45%), indicating that this biological control agent would remain effective for controlling the weed in CO_2-enriched environment (Shabbir et al. 2015). Disease pathogens are likely to perform more effectively as biological control agents to manage the weeds under CO_2-enriched environment. For example, it was observed that winter rust (*Puccinia abrupta* var. *partheniicola*) infected plants of *P. hysterophorus* exhibited decreased photosynthesis and water use efficiency under elevated CO_2 levels (Shabbir et al. 2014).

Difficulties Arising from Ambient Condition Warming

Environmental conditions like temperature and relative humidity at herbicide application time can change the herbicide efficacy either by changing absorption or translocation by altering the fluidity of cuticle and plasma membranes (Sharma and Singh 2001; Johnson and Young 2002; Ramsey et al. 2005). High temperatures may also increase the metabolism rate of herbicide within the plant system (Johnson and Young 2002), and an increased temperature level may cause stress to plants and alter the leaf cuticular composition and absorption of the herbicide (Ziska and Bunce 2006). Warming of moist soil causes more volatilization losses of surface-applied chloroacetanilide herbicides (Beestman and Deming 1974) and degradation of imidazolinone herbicides (Mcdowell et al. 1997). Indeed, herbicide degradation is also faster under high soil temperature conditions, thus reducing the herbicide availability from the soil. In these circumstances, the control of *Alternanthera philoxeroides* was reduced from 92% to 80% when penoxsulam was applied at temperatures of 21/11 °C and 27/18 °C, respectively (Willingham et al. 2008). Warming may also reduce herbicide persistence in the soil and the efficacy (Bailey 2004) of soil-applied herbicide.

Dealing with Problems Caused by Water Stress

In the future, water scarcity in Asia may arise from both physical and economic causes by 2025 (Tuong and Bouman 2003), and in Africa, rainfed agriculture systems will be more vulnerable to drought (Rodenburg et al. 2011), and these are concerning scenarios for cultures which survive on agricultural output. For example, it is known that C_4 plants are dominant under dryland soils and C_3 plants are dominant in submerged soils, and this difference may be the reason that submergence protects Asian rice crops from competition by weeds of C_4 photosynthetic pathway (Matsunaka 1983). In rainfed areas, drought periods may cause thick cuticle development on leaves, increased leaf pubescence and closure of stomata (Patterson et al. 1999) or reduced leaf area and change in leaf orientation (Zhou et al. 2007). These changes will severely decrease the ability for herbicide uptake and thus damage crop production, particularly in Africa. Generally, in dryland agriculture, crop-weed competition will become more severe and timely management of weeds will attain importance when attempting to realize crop potential yields (Le Houerou 1996).

Some Pointers Towards Future Crop Production Management

Innovative Crop Breeding Efforts

In the future, it will be essential to develop a new suite of crops with efficient germplasms which will exhibit desired traits like drought/heat tolerance and weed-smothering potential. In parallel with this development, there will be the potential for adaptation of advanced agricultural technologies which will mitigate the amount of GHG emissions with their adoption. In addition, selection of crop cultivars which have high-yield potential under a changing climate scenario and have weed-suppressive potential (or compete well with weeds) is of major importance (Korres et al. 2016). Of continuing concern, however, is that the greater innate genetic diversity present in weedy lines will allow them to develop adaptability to rising temperatures and increased atmospheric CO_2 levels. On the other hand, although crops have been bred and selected for high yields in different climate scenarios, there may be a future novel source of resistance to biotic and abiotic stresses which can be extracted from weedy lines, and this may be used to develop new varieties of hardier crop varieties (Ziska and McClung 2008).

Advanced breeding techniques utilize the continuously adapting germplasm of current crops to ongoing climate change and are evaluating the emergent crops at multiple locations of varied environments (Barnaby and Ziska 2012). Of high priority is to develop strains with increased drought tolerance, which will reduce irrigation water requirements and pumping energy. Such efforts will clearly contribute to the possibility of achieving sustainable crop yields in areas receiving less rainfall. Breeding crops for evolutionary traits, such as increased nitrogen use efficiency, should be a top priority since this will improve the GHG profile of crop production on the basis of reduction in nitrogen oxide emissions. Also, under enhanced CO_2 levels, the nutritive value of crop plants will be needed to be re-examined, as plants will have more starch but less protein content and will evidence reduced digestibility (Taub et al. 2008).

Conservation Agriculture Systems

Conservation agriculture is an adaptation and mitigation strategy which is currently being practiced in many countries as a response to climate change conditions. To address the obvious difficulties introduced by the weather effects and other variations to farming practice, systems referred to as climate-resilient agriculture, zero tillage, or conservation agriculture have been recommended. One mitigation strategy has been to sequester additional carbon in the soil (Lal et al. 1998), either as buried gaseous CO_2 or as mulched material. However, the success of conservation agriculture will critically depend upon successful chemical weed control. The cultivation of herbicide-tolerant crops has facilitated conservation agriculture systems

(Brookes and Barfoot 2008), and conservation tillage techniques have reduced GHGs emissions, especially CO_2 due to reduced fuel use. Carbon sequestration using crop residue management techniques have been seen to be useful, although nitrogen oxides emissions have been seen to rise with the introduction of CA systems due to more soil moisture and less porosity (Holland 2004). Nevertheless, adherence to the three principles of conservation agriculture, namely, zero tillage, crop rotation, and retention of residues on the surface, and use of cover crops or mulches will help in maintaining adequate weed control when used in conjunction with reduced chemical use.

Improved Agronomic Practices for Chemical Weed Control

The use of improved varieties of crops, coupled with refined agrotechnology practices, have resulted in the achievement of higher crop productivities. However, it is anticipated that to attain a state of sustainable crop production, more precision agricultural techniques will be needed to be adopted. These will include infrastructure such as modern planters, sprayers, and other harvest equipment. In particular, herbicide spray technology will have to be refined in term of time of spray and amount of herbicide delivered. This will be in concert with a need for integrated weed management methods which will be required to be developed in order to (i) prevent ingress of weeds into new areas, (ii) interfere with weed population reproduction, (iii) and to prevent the further dispersal and establishment of weeds.

Weeds Which Cause Allergic and Toxic Reactions

As a final remark, we note that the impact of climate change on growth, phenology, and pollen production of weeds with known allergic reactions to human systems and to other economically important plants needs to be studied in depth (Ziska et al. 2009). It is already known that respiratory allergies are caused by weeds like *Ambrosia* spp., *Betula* spp., and *P. hysterophorus* and contact dermatitis has been shown to be caused by *Toxicodendron* spp./*Rhus* spp. (Ziska 2003b). Such weeds will present immediate problems for agricultural workers, but there can also be wind-borne material which will create wider difficulties for populations near crop-growing situations. In this respect, both the increase in production of pollen under climate stress conditions, together with the attendant increase in strength of wind gusts will combine to exacerbate these problems. In addition, some commonly encountered weeds, such as *Ricinus communis* and *Rheum rhabarbarum*, are directly toxic to plants and will therefore be responsible for unavoidable crop mortality. Unless immediate steps are taken to curtail the spread of such species, significant economic implications will arise – coming both from acute weed control actions and significant crop yield decreases.

Conclusions

It is recognized that weeds are extremely dynamic in nature and thus will most likely be able to acclimatize effortlessly to all forms of climate change. Indeed, there has been some research suggesting that weeds are able to aggressively flourish in higher CO_2 atmospheres, warmer weather, and increased water stress conditions. However, with respect to competition in practical cropping situations, the picture is not yet clear, since the limited number of research studies have mainly dealt with single crop/weed pairs, based on an investigation of the efficacy of C_3 vs a C_4 photosynthetic pathway under controlled environments. Notwithstanding these valuable contributions, in a field situation, crop plants are inevitably surrounded by an assemblage of C_3 and C_4 plants, and it is known that there is a considerable variation in the growth response of competing plants to climate change. Given the wisdom of the old adage 'Forewarned is forearmed', it seems prudent that, faced with the projected changes in multiple climatic parameters and soil resources, a multifactor approach will be needed to study the effects of climate change on a variety of weed growth and management situations and on an equally large set of crop-weed interference mechanisms. In addition to this type of programme, because the nature and severity of the outcomes of concurrent changes in CO_2 and other physical parameters on crop-weed competition are still unclear, there is urgency for a deep understanding of the interactive effects of climate change parameters. Data regarding the interactions of air temperature, CO_2, ozone, precipitation pattern, soil nutrition and water availability with respect to plant growth, are paramount. At a practical level, studies of a range of crops and weed species grown in monoculture and mixed-species communities will be required to allow a systematic development of crop management strategies to emerge.

Emphasising the complexity of this task, we note that the differential response of C_3 and C_4 plants to elevated CO_2 and ambient warming will have important implications on the future distribution and spread of weed species. In addition, the genus shift in weed flora from annuals to perennials in continuous zero-till fields, coupled with the reduced efficacy of many herbicides used for weed control, could significantly negate the beneficial effects of zero tillage on carbon sequestration. Indeed, for the development of sustainable and effective weed management with herbicides, new strategies will have to be adopted to address weed infestations under changing climate scenarios. A number of tentative suggestions have been made in this regard including the increase of the active ingredient concentration in a herbicide, increasing the number of herbicide applications, or moving to the addition of surfactants to allow more penetration of applied herbicides.

However, it is feared that this over-reliance on herbicides will add an extra burden to the management task with respect to both the economics of production and the surrounding environment. It will also significantly increase the risk of accelerated herbicide resistance evolution in weeds. In addition, it is anticipated that those weeds with asexual/vegetative propagation means, such as perennial weeds, will pose a further major hurdle in sustainable crop production as these weeds become progressively harder to control with chemical and mechanical weed control methods in a higher CO_2 world. This problem will be exacerbated by the expected increase in gene flow between cultivated crops and their wild relatives under a climate change scenario, which will resoundingly decide the future of herbicide-tolerant crops. Finally, we assert that there is a compelling requirement to conduct extensive research to assess the potential consequences of climate change on the efficacy of different herbicide families, in order to allow considered selection of herbicide types in specific climatic situations. In this respect, there seems to be no alternative but to plan and conduct long-term field experiments with multiple climate variables to prepare crop managers to adequately face the multiple challenges of weeds for economic crop production.

References

Adams RM, Hurd BH, Lenhart S, Leary N (1998) Effects of global climatic change on agriculture: an interpretative review. Climate Res 11:19–30

Ainsworth EA, Long SP (2005) What have we learned from 15 years of Free-Air CO_2 Enrichment (FACE)? A meta-analysis review of the responses of photosynthesis, canopy properties and plant production to rising CO_2. New Phytol 165:351–371

Ainsworth EA, Davey PA, Bernacchi CJ, Dermody OC, Heaton EA, Moore DJ, Morgan PB, Naidu SL, Yoo Ra HH, Zhu XG, Curtis PS, Long SP (2002) A meta-analysis of elevated [CO_2] effects on soybean (*Glycine max*) physiology, growth and yield. Glob Chang Biol 8:695–709

Akita S, Moss DN (1972) Differential stomatal response between C_3 and C_4 species to atmospheric CO_2 concentration and light. Crop Sci 12:789–793

Alberto AM, Ziska LH, Cervancia CR, Manalo PA (1996) The influence of increasing carbon dioxide and temperature on competitive interactions between a C_3 crop and a C_4 weed. Aust J Plant Physiol 23:795–802

Anten NP, Hirose T, Onoda Y, Kinugasa T, Kim YH, Okada M (2003) Elevated $CO_2$and nitrogen availability have interactive effects on canopy carbon gain in rice. New Phytol 161:459–471

Archambault DJ (2007) Efficacy of herbicides under elevated temperature and CO_2. In: Newton PCD, Carran RA, Edwards GR, Niklaus PA (eds) Agroecosystems in a changing climate. CRC Press, Boca Raton, pp 272–279

Bailey SW (2004) Climate change and decreasing herbicide persistence. Pest Manag Sci 60:158–162

Baker HG (1965) Characteristics and mode of origin of weeds. In: Baker HG, Stebbins GL (eds) The genetics of colonizing species. Academic Press, New York, pp 147–168

Baldocchi D, Wong S (2008) Accumulated winter chill is decreasing in the fruit growing regions of California. Clim Change 87(1):153–166

Barnaby JY, Ziska LH (2012) Plant responses to elevated CO_2. In: Encyclopedia of Life Sciences (eLS). Wiley, p e23718

Batts GR, Morison JIL, Ellis RH, Hadley P, Wheeler TR (1997) Effects of CO_2 and temperature on growth and yield of crops of winter wheat over four seasons. Eur J Agron 7:43–52

Bazzaz FA, Carlson MR (1984) The response of plants to elevated $CO_{2.}$ I Competition among the assemblage of annuals at two levels of soil moisture. Oecologia 62:196–198

Bazzaz FA, Garbutt K, Reekie EG, Williams WE (1989) Using growth analysis to interpret competition between a C_3 and C_4 annual under ambient and elevated CO_2. Oecologia 79:223–235

Beestman GB, Deming JM (1974) Dissipation of acetanilide herbicides from soils. Agron J 66(2):308–311

Belote RT, Weltzin JF, Norby RJ (2003) Response of an understory plant community to elevated [CO_2] depends on differential responses of dominant invasive species and is mediated by soil water availability. New Phytol 161:827–835

Blank RR, White R, Ziska LH (2006) Combustion properties of *Bromus tectorum* L.: influence of ecotype and growth under four CO_2 concentrations. Int J Wildland Fire 15:227–236

Bleier JS, Jackson RD (2007) Manipulating the quantity, quality and manner of C addition to reduce soil inorganic N and increase C_4:C_3 grass biomass. Restor Ecol 15:688–695

Blum A (1996) Crop responses to drought and the interpretation of adaptation. Plant Growth Regul 20:135–148

Blumenthal D, Chimner RA, Welker JM, Morgan JA (2008) Increased snow facilitates plant invasion in mixed grass prairie. New Phytol 179:440–448

Booth BD, Swanton CJ (2002) Assembly theory applied to weed communities. Weed Sci 50:2–13

Bowes G (1996) Photosynthesis responses to changing atmospheric carbon dioxide concentration. In: Baker NR (ed) Photosynthesis and the environment. Kluwer Publ., Dordrecht, pp 87–407

Boyce DG, Lewis MR, Worm B (2010) Global phytoplankton decline over the past century. Nature 466:591

Boydston RA, Mojtahedi H, Crosslin JM, Brown CR, Anderson T (2008) Effect of hairy night shade (*Solanum sarrachoides*) presence on potato nematodes, diseases, and insect pests. Weed Sci 56:151–154

Bridges DC (1992) Crop losses due to weeds in the United States. Weed Science Society of America, Champaign, 403p

Brookes G, Barfoot P (2008) Global impact of biotech crops: socioeconomic and environmental effects, 1996–2006. AgBioforum 11:21–38

Bunce JA (1995) Long-term growth of alfalfa and orchard grass plots at elevated carbon dioxide. J Biogeogr 22:341–348

Bunce JA, Ziska LH (2000) Crop ecosystem responses to climatic change: crop/weed interactions. In: Reddy KR, Hodges HF (eds) Climate change and global crop productivity. CABI Publishing, New York, pp 333–352

Burney JA, Davis SJ, Lobell DB (2010) Greenhouse gas mitigation by agricultural intensification. Proc Natl Acad Sci U S A 10:12052–12057

Canadell JG, LeQuéré C, Raupach MR, Field CB, Buitenhuis ET, Clais P, Conway TJ, Gillett NP, Houghton RA, Marland G (2007) Contributions to accelerating atmospheric growth from economic activity, carbon intensity and efficiency of natural sinks. Proc Natl Acad Sci U S A 104:18866–18870

Carlson HL, Hill JE (1985) Wild oat (*Avena fatua*) competition with spring wheat: effects of nitrogen fertilization. Weed Sci 34:29–33

Chamberlain J, Gittens A (2004) Parthenium weed-management: challenges, opportunities and strategies. Parthenium Action Group. The State of Queensland Department of Natural Resources, Mines and Energy, Brisbane, p 82

Chauhan BS, Ramesh K (2015) Weed regimes in agro-ecosystems in the changing climate scenario – A review. Ind J Agron 60(4):479–484

Chauhan BS, Prabhjyot-Kaur MG, Randhawa RK, Singh H, Kang MS (2014) Global warming and its possible impacts on agriculture in India. Adv Agron 123:65–121

Clements DR, Ditommaso A (2011) Climate change and weed adaptation: can evolution of invasive plants lead to greater range expansion than forecasted? Weed Res 51:227–240

Coble HD, Williams FM, Ritter RL (1981) Common ragweed (*Ambrosia artemissifolia*) interference in soybean (*Glycine max*). Weed Sci 29:339–342

Conroy JP, Seneweera S, Basra AS, Rogers G, Nissenwooller B (1994) Influence of rising CO_2 concentrations and temperature on growth, yield and grain quality of cereal crops. Aust J Plant Physiol 21:741–758

Cushman RM (1999) Global climatic-change modeling and monitoring. In: Alexander DE, Fairbridge RW (eds) Encyclopedia of environmental science. Kluwer Academic Publishers, Dordrecht, pp 291–298

Dekker J, Duke OS (1995) Herbicide resistant field crops. Adv Agron 54: 69–116
Délye C (2013) Unravelling the genetic bases of non-target-site based resistance (NTSR) to herbicides: a major challenge for weed science in the forthcoming decade. Pest Manag Sci 69:176–187
Délye C, Jasieniuk M, Le Corre V (2013) Deciphering the evolution of herbicide resistance in weeds. Trends Genet 29:649–658
Dikšaityte A, Januškaitiene I and Juknys R (2014) Impact of elevated atmospheric CO_2 concentration on pea and white melilot at three levels of nitrogen fertilization. In: The 9th international conference "Environmental Engineering". Section environmental protection. Selected papers. Article number: enviro.2014.015
Donald WW (1990) Management and control of Canada thistle. Rev Weed Sci 5:193–250
Dyson T (2001) World food trends: a neo-Malthusian prospect? Proc Am Philosophical Soc 145:438–455
Edwards GE, Huber S (1981) The C_4 pathway. In: Hatch MD, Boardman NK (eds) The biochemistry of plants. Academic, New York, pp 238–281
Ellstrand NC, Prentice HC, Hancock JF (1999) Gene flow and introgression from domesticated plants into their wild relatives. Annu Rev Ecol Syst 30:539–563
Erickson JE, Megonigal JP, Peresta G, Drake BG (2007) Salinity and sea level mediate elevated CO_2 effects on C_3-C_4 plant interactions and tissue nitrogen in a Chesapeake Bay tidal wetland. Glob Chang Biol 13:202–215
Fantke P, Gillespie BW, Juraske R, Jolliet O (2014) Estimating half-lives for pesticide dissipation from plants. Environ Sci Technol 48:8588–8602
FAOSTAT (2018) World summit on food security. Food and agriculture organization of the united nations-statistics division. http://www.fao.org/wsfs/world-summit/en. Accessed 25 Feb 2018
Fernando N, Panozzo J, Tausz M, Norton R, Fitzgerald G, Khan A, Seneweera S (2015) Rising CO_2 concentration altered wheat grain proteome and flour rheological characteristics. Food Chem 170:448–454
Fernando N, Manalil S, Florentine SK, Chauhan BS, Seneweera S (2016) Glyphosate resistance of C_3 and C_4 weeds under rising atmospheric CO_2. Front Plant Sci 7:910
Flint EP, Patterson DT, Beyers JL (1983) Interference and temperature effects on growth of cotton (*Gossypium hirsutum*), spurred anoda (*Anoda cristata*), and velvetleaf (*Abutilon theophrasti*). Weed Sci 31:892–898
Fried G, Petit S, Rebound X (2010) A specialist-generalist classification of the arable flora and its response to changes in agricultural practices. Bio Med Central Ecol 10:1–11
Froud-Williams RJ (1996) Weeds and climate change: implications for their ecology and control. Asp Appl Biol 45:187–196
Fuhrer J (2003) Agroecosystem responses to combinations of elevated CO_2, ozone, and global climate change. Agric Ecosyst Environ 97:1–20
Gains TA, Shaner DL, Ward SM, Leach JE, Preston C, Westra P (2011) Mechanism of resistance of evolved glyphosate-resistant palmer amaranth (*Amaranthus palmeri*). J Agric Food Chem 59:5886–5889
Ge X, D'Avignon D, Ackerman JJH, Duncan B, Spaur MB, Sammons RD (2011) Glyphosate-resistant horseweed made sensitive to glyphosate: low-temperature suppression of glyphosate vacuolar sequestration revealed by 31P NMR. Pest Manag Sci 67:1215–1221
Gealy DR, Mitten DH, Rutger JN (2003) Gene flow between red rice (*Oryza sativa*) and herbicide resistant rice (*O. sativa*): implications for weed management. Weed Technol 17:627–645
Gianessi LP (2013) The increasing importance of herbicides in worldwide crop production. Pest Manag Sci 69:1099–1105
Gibson KD, Johnson WG, Hillger DE (2005) Farmer perceptions of problematic corn and soybean weeds in Indiana. Weed Technol 19:1065–1070
Griffin JG, Thomas GR, Mason-Pharr D (2004) Heat and drought influence photosynthesis, water relations, and soluble carbohydrates of two ecotypes of redbud (*Cercis canadensis*). J Am Soc Hort Sci 129(4):497–502

Gustafson DI (2011) Climate change: a crop protection challenge for the twenty-first century. Pest Manag Sci 67:691–696

Hall AE, Allen LH Jr (1993) Designing cultivars for the climatic conditions of the next century. In: Buxton DR, Shibles R, Forsberg RA, Blad BL, Asay KH, Paulsen GM, Wilson RF (eds) International crop sci. Crop Science Society of America, Madison, pp 291–297

Han H, Yu Q, Cawthray GR, Powles SB (2013) Enhanced herbicide metabolism induced by 2, 4-D in herbicide susceptible *Lolium rigidum* provides protection against diclofop-methyl. Pest Manag Sci 69:996–1000

Hanzlik K, Gerowitt B (2012) Occurrence and distribution of important weed species in German winter oilseed rape fields. J Plant Dis Prot 119:107–120

Hasanuzzaman M, Nahar K, Alam M, Roychowdhury R, Fujita M (2013) Physiological, biochemical and molecular mechanisms of heat stress tolerance in plants. Int J Mol Sci 14:9643–9684

Hatch M (1987) C_4 photosynthesis: a unique blend of modified biochemistry, anatomy and ultrastructure. Biochem Biophys Acta 895:81–106

Heap I (2018) The international survey of herbicide resistant weeds. http://www.weedscience.com. Accessed 25 Feb 2018

Hedden P (2003) The genes of the green revolution. Trends Genet 19:5–9

Högy P, Wieser H, Koehler P, Schwadorf K, Breuer J, Franzaring J, Muntifering R, Fangmeier A (2009) Effects of elevated CO_2 on grain yield and quality of wheat: results from a 3-year free-air CO_2 enrichment experiment. Plant Biol 11:60–69

Holland JM (2004) The environmental consequences of adopting conservation tillage in Europe: reviewing the evidence. Agric Ecosyst Environ 103:1–25

HRAC (2018) Classification of herbicides according to mode of action. http://wwwhracglobalcom Accessed 25 Feb 2018

IPCC (2007) Climate change 2007: impacts, adaptation and vulnerability. In: Parry ML, Canziani OF, Palutikof JP, van der Linden PJ, Hanson CE (eds) Contribution of working group II to the fourth assessment report of the intergovernmental panel on climate change. Cambridge University Press, Cambridge, 976

Johnson BC, Young BG (2002) Influence of temperature and relative humidity on the foliar activity of mesotrione. Weed Sci 50(2):157–161

Jump AS, Peñuelas J (2005) Running and stand still: adaptation and the response of plants to rapid climate change. Ecol Lett 8:1010–1020

Kathiresan RM (2017) Anticipatory research in weed science to mitigate climate change impact. In: Proceedings of biennial conference of the Indian Society of Weed Science on "Doubling farmers' income by 2022: The role of weed science", held during 1–3 March at MPUA&T, Udaipur, India, pp 11

Kimball BA, Kobayashi K, Bindi M (2002) Response of agricultural crops to free-air CO_2 –enrichment. Adv Agron 77: 293–368

Knapp AK, Hamerlyn CK, Owensby CE (1993) Photosynthetic and water relations response to elevated CO_2 in the C_4 grass, *Andropogon gerardii*. Int J Plant Sci 154:459–466

Korres NE (2005) Encyclopaedic dictionary of weed science: theory and digest. Intercept Andover, Paris, 695

Korres NE, Norsworthy JK, Bagavathiannan MV, Mavromoustaki A (2015) Distribution of arable weed populations along eastern Arkansas Mississippi Delta roadsides; factors affecting weed occurrence. Weed Technol 29(3):596–604

Korres NE, Norsworthy JK, Tehranchian P, Gitsopoulos TK, Loka DA, Oosterhuis DM, Gealy DR, Moss SR, Burgos NR, Miller MR, Palhano M (2016) Cultivars to face climate change effects on crops and weeds: a review. Agron Sustain Dev 36:12

Kraehmer H, Almsick VA, Beffa R, Dietrich H, Eckes P, Hacker E, Hain R, Strek HJ, Stuebler H, Willms L (2014) Herbicides as weed control agents: state of the art: II. Recent achievements. Plant Physiol 166:1132–1148

Kudsk P, Kristensen JL (1992) Effect of environmental factors on herbicide performance. Proceedings of the First International Weed Control Congress, Melbourne, pp 173–186

Lal R, Kimble JM, Follett RF, Cole CV (1998) The potential of U.S. cropland to sequester carbon and mitigate the greenhouse effect. Ann Arbor Press, Chelsea, 128

Lara MV, Andreo CS (2011) C_4 plants adaptation to high levels of CO_2 and to drought environments. In: Shankar A (ed) Abiotic stress in plants – mechanisms and adaptations. InTech, pp 415–428

Lasat MM, DiTomaso JM, Hart JJ, Kochian LV (1996) Resistance to paraquat in *Hordeum glaucum* is temperature dependent and not associated with enhanced apoplasmic binding. Weed Res 36:303–309

Le Houerou HN (1996) Climate changes, drought and desertification. J Arid Environ 34:133–185

Leakey ADB, Ainsworth EA, Bernacchi CJ, Rogers A, Long SP, Ort DB (2009) Elevated CO_2 effects on plant carbon, nitrogen, and water relations: six important lessons from FACE. J Exp Bot 60:2859–2876

Lee JS (2011) Combined effect of elevated CO_2 and temperature on the growth and phenology of two annual C_3 and C_4 weedy species. Agric Ecosyst Environ 140:484–491

Leegood RC (2002) C_4 photosynthesis: principles of CO_2 concentration and prospects for its introduction into C_3 plants. J Exp Bot 53:581–590

Long SP (1999) Environmental responses. In: Sage RF, Monson RK (eds) C_4 plant biology. Academic Press, San Diego, pp 215–249

Long SP, Ainsworth EA, Rogers A, Ort DR (2004) Rising atmospheric carbon dioxide: plants FACE the future. Annu Rev Plant Biol 55:591–628

Long SP, Ainsworth EA, Leakey ADB, Nösberger J, Ort DR (2006) Food for thought: lower-than-expected crop yield stimulation with rising CO_2 concentration. Science 312:1918–1921

Mahan JR, Dotray PA, Light GG, Dawson KR (2006) Thermal dependence of bioengineered glufosinate tolerance in cotton. Weed Sci 54:1–5

Manea A, Leishman MR, Downey PO (2011) Exotic C_4 grasses have increased tolerance to glyphosate under elevated carbon dioxide. Weed Sci 59:28–36

Martinez-Ghersa MA, Ghersa CM, Satorre EH (2000) Co-evolution of agricultural systems and their weed companions: implications for research. Field Crop Res 67:181–190

Matsunaka S (1983) Evolution of rice weed control practices and research: world perspective. In: Weed control in rice. IRRI, Manila, pp 5–18

Matzrafi M, Seiwert B, Reemtsma T, Rubin B, Peleg Z (2016) Climate change increases the risk of herbicide-resistant weeds due to enhanced detoxification. Planta 244: 1217–1227

McDonald A, Riha S, DiTommaso A, DeGaetano A (2009) Climate change and the geography of weed damage: analysis of U.S. maize systems suggests the potential for significant range transformations. Agric Ecosyst Environ 130:131–140

Mcdowell RW, Condron LM, Main BE, Dastgheib F (1997) Dissipation of imazapyr, flumetsulam and thifensulfuron in soil. Weed Res 37(6):381–389

McFadden JJ, Frear CS, Mansager ER (1989) Aryl hydroxylation of diclofop by a cytochrome P450 dependent monooxygenase from wheat. Pestic Biochem Physiol 34:92–100

Meissle M, Mouron P, Bigler MT, Pons X, Vasileiadis VP, Otto S, Antichi D, Kiss J, Pálinkás Z, Dorner Z, Weide RVD, Groten J, Czembor E, Adamczyk J, Thibord JB, Melander B, Nielsen GC, Poulsen RT, Zimmermann O, Verschwele A, Oldenburg E (2010) Pests, pesticide use and alternative options in European maize production: current status and future prospects. J Appl Entomol 134:357–375

Milakovic I, Fiedler K, Karrer G (2014) Management of roadside populations of invasive *Ambrosia artemisiifolia* by mowing. Weed Res 54:256–264

Miri HR, Rastegar A, Bagheri AR (2012) The impact of elevated CO_2 on growth and competitiveness of C_3 and C_4 crops and weeds. Euro J Exp Bio 2(4):1144–1150

Moore BD, Franceschi VR, Cheng SH, Wu J, Ku MSB (1987) Photosynthetic characteristics of the C_3-C_4 intermediate *Parthenium hysterophorus*. Plant Physiol 85:984–989

Morrison JI, Gifford RM (1984) Plant growth and water use with limited supply in high CO_2 concentrations. I. Leaf area, water use and transpiration. Aust J Plant Physiol 11:361–374

Mortensen DA, Coble HD (1989) The influence of soil water content on common cocklebur (*Xanthium strumarium*) interference with soybean (*Glycine max*). Weed Sci 37:76–83
Mpelasoka F, Hennessy K, Jones R, Bates B (2008) Comparison of suitable drought indices for climate change impacts assessment over Australia towards resource management. Int J Climatol 28:1283–1292
Naidu VSGR, Paroha S (2008) Growth and biomass partitioning in two weed species *Parthenium hysterophorus* (C_3) and *Amaranthus viridis* (C_4) under elevated CO_2. Eco Env Cons 14(4):9–12
Naidu VSGR, Varshney JG (2010) Interactive effect of elevated CO_2, drought and weed competition on carbon isotope discrimination (Δ13C) in wheat (*Triticum aestivum*) leaves. Indian J Agric Sci 81(11):1026–1029
Noctor G, Mhamdi A, Chaouch S, Han Y, Neukermans J, Marquez-Garcia B, Queval G, Foyer CH (2012) Glutathione in plants: an integrated overview. Plant Cell Environ 35:454–484
O'Donnell CC, Adkins SW (2001) Wild oat and climate change: the effect of CO_2 concentration, temperature, and water deficit on the growth and development of wild oat in monoculture. Weed Sci 49:694–702
Oerke EC (2006) Crop losses to pests. J Agric Sci 144:31–43
Ozturk M, Rehder H, Zeigler H (1981) Biomass production of C_3 and C_4 plant species in pure and mixed culture with different water supply. Oecologia (Berl) 50:73–81
Pandey DK, Palni LMS, Joshi SC (2003) Growth, reproduction, and photosynthesis of ragweed parthenium (*Parthenium hysterophorus*). Weed Sci 51:191–201
Parmesan C (1996) Climate and species' range. Nature 382:765–766
Patra AK, Kumar M, Justin GK (2012) Potential impact of climate change on soil fertility and fertilizer use in India. Ind J Fert 8(11):42–55
Patterson DT (1985) Comparative eco-physiology of weeds and crops. In: Duke SO (ed) Weed physiology. CRC Press, Boca Raton, pp 101–129
Patterson DT (1986) Responses of soybean (*Glycine max*) and three C_4 grass weeds to CO_2 enrichment during drought. Weed Sci 34:203–210
Patterson DT (1993) Implications of global climate change for impact of weeds, insects and plant diseases. In: Buxton DR (ed) International crop science I. Crop Science Society of America, Madison, pp 273–280
Patterson DT (1995a) Effects of environmental stress on weed/crop interactions. Weed Sci 43:483–490
Patterson DT (1995b) Weeds in a changing climate. Weed Sci 43:685–701
Patterson DT, Flint EP (1980) Potential effects of global atmospheric CO_2 enrichment on the growth and competitiveness of C_3 and C_4 weed and crop plants. Weed Sci 28:70–75
Patterson DT, Meyer CR, Flint EP, Quimby PC Jr (1979) Temperature responses and potential distribution of itchgrass (*Rottboellia exaltata*) in the United States. Weed Sci 27:77–82
Patterson DT, Flint EP, Dickens R (1980) Effects of temperature, photoperiod, and population source on the growth of cogongrass (*Imperata cylindrica*). Weed Sci 28:205–509
Patterson DT, Musser RL, Flint EP, Eplee RE (1982) Temperature responses and potential for spread of witchweed (*Striga lutea*) in the United States. Weed Sci 30:87–93
Patterson DT, Flint EP, Beyers JL (1984) Effectsof CO_2 enrichment on competition between a C_4 weed and a C_3 crop. Weed Sci 32:101–105
Patterson DT, Russell AE, Mortensen DA, Coffin RD, Flint EP (1986) Effects of temperature and photoperiod on Texas panicum (*Panicum texanum*) and wild proso millet (*Panicum miliaceum*). Weed Sci 34:876–882
Patterson DT, Highsmith MT, Flint EP (1988) Effects of temperatutre and CO_2 concentration on the growth of cotton (*Gossypium hirsutum*), spurred anoda (*Anoda cristata*), and velvetleaf (*Abutilon theophrasti*). Weed Sci 36:751–757
Patterson DT, Westbrook JK, Joyce RJV, Lingren PD, Rogasik J (1999) Weeds, insects and diseases. Clim Change 43:711–727
Pearman PB, Guisan A, Broennimann O, Randin CF (2008) Niche dynamics in space and time. Trends Ecol Evol 23:149–158

Peters K, Breitsameter L, Gerowitt B (2014) Impact of climate change on weeds in agriculture: a review. Agron Sustain Dev 34:707–721

Portis AR Jr, Parry MAJ (2007) Discoveries in Rubisco (ribulose 1,5-biphosphate carboxylase/oxygenase): a historical perspective. Photosynth Res 94:121–143

Ramsey RJL, Stephenson GR, Hall JC (2005) A review of the effects of humidity, humectants and surfactant composition on the absorption and efficacy of highly water-soluble herbicides. Pest Biochem Physiol 82(2):162–175

Rivers J, Warthmann N, Pogson B, Borevitz J (2015) Genomic breeding for food, environment and livelihoods. Food Security 7:375–382

Robinson EA, Ryan GD, Newman JA (2012) Meta-analytical review of the effects of elevated CO_2 on plant-arthropod interactions highlights the importance of interacting environmental and biological variables. New Phytol 194:321–336

Robinson MA, Letarte J, Cowbrough MJ, Sikkema PH, Tardif FJ (2015) Winter wheat (*Triticum aestivum* L.) response to herbicides as affected by application timing and temperature. Can J Plant Sci 95:325–333

Rodenburg J, Meinke H, Johnson DE (2011) Challenges for weed management in African rice systems in a changing climate. J Agric Sci 149:427–435

Rogers HH, Runion GB, Prior SA, Price AJ, Torbert HA, Gjerstad DH (2008) Effects of elevated atmospheric CO_2 on invasive plants: comparison of purple and yellow nutsedge (*Cyperus rotundus* L. and *C. esculentus* L.). J Environ Qual 37:395–400

Rogers A, Ainsworth EA, Leakey ADB (2009) Will elevated CO_2 concentration amplify the benefits of nitrogen fixation in legumes? Plant Physiol 151:1009–1016

Rosenzweig C, Hillel D (1998) Effects on weeds, insects and diseases. In: Rosenzweig C, Hillel D (eds) Climate change and the global harvest. Oxford University Press, Oxford, pp 101–122

Sage RF, Pearcy RW (1987) The nitrogen use efficiency of C_3 and C_4 plants. I. Leaf nitrogen, growth and biomass partitioning in *Chenopodium album* and *Amaranthus retroflexus*. Plant Physiol 84:954–958

Sage RF, Christin PA, Edwards EA (2011) The C_4 plant lineages of planet Earth. J Exp Bot 62:155–169

Sasek TW, Strain BR (1990) Implications of atmospheric CO2 enrichment and climatic change for the geographical distribution of two introduced vines in the USA. Clim Change 16: 31–51

Schlenker W, Roberts MJ (2009) Nonlinear temperature effects indicate severe damages to US crop yields under climate change. Proc Natl Acad Sci U S A 106:15594–15598

Seneweera S, Makino A, Mae A, Basra AS (2005) Response of rice to p[CO_2] enrichment: the relationship between photosynthesis and nitrogen metabolism. J Crop Improv 13:31–53

Shabbir A, Dhileepan K, Khan N, Adkins SW (2014) Weed-pathogen interactions and elevated CO_2: growth changes in favour of the biological control agent. Weed Res 54:217–222

Shabbir A, Dhileepan K, Zalucki MP, Adkins SW (2015) The response of an invasive weed and its biological control agent under a changing climate of CO_2 enrichment: management challenges for the future. In: Proceedings of 25th Asian-Pacific Weed Science Society Conference on "Weed Science for Sustainable Agriculture, Environment and Biodiversity", Hyderabad, India during 13–16 October, p 33

Sharma SD, Singh M (2001) Environmental factors affecting absorption and bio-efficacy of glyphosate in Florida beggarweed (*Desmodium tortuosum*). Crop Prot 20:511–516

Shivrain VK, Burgos NR, Anders MM, Rajguru SM, Moore J, Sales MA (2007) Gene flow between Clearfield™ rice and red rice. Crop Prot 26:349–356

Shivrain VK, Burgos NR, Sales MA, Mauromoustakos A, Gealy DR, Smith KL, Black HL, Jia M (2009) Factors affecting the outcrossing rate between Clearfield™ rice and red rice (*Oryza sativa*). Weed Sci 57:394–403

Singh RP, Singh RK, Singh MK (2011) Impact of climate and carbon dioxide change on weeds and their management. Indian J Weed Sci 43(1–2):1–11

Sionit N, Hellmers H, Strain BR (1980) Growth and yield of wheat under CO_2 enrichment and water stress. Crop Sci 20:687–690

Song L, Wu J, Changhan L, Furong L, Peng S, Chen B (2009) Different responses of invasive and native species to elevated CO_2 concentration. Acta Oecol 35:128–135

Stitt M (1991) Rising CO_2 levels and their potential significance for carbon flow in photosynthetic cells. Plant Cell Environ 14:741–762

Storkey J, Cussans JW (2000) Relationship between temperature and the early growth of *Triticum aestivum* and three weed species. Weed Sci 48:467–473

Stratonovitch P, Storkey J, Semenov AA (2012) A process-based approach to modeling impacts of climate change on the damage niche of an agricultural weed. Glob Chang Biol 18:2071–2080

Tanaka R, Koike F (2011) Prediction of species composition of plant communities in a rural landscape based on species traits. Ecol Res 26:27–36

Taub DR, Miller B, Allen H (2008) Effects of elevated CO_2 on the protein concentration of food crops: a meta-analysis. Glob Chang Biol 14:565–575

Treharne K (1989) The implications of the 'greenhouse effect' for fertilizers and agrochemicals. In: Bennet RC (ed) The greenhouse effect and UK agriculture, CAS Paper 19. Univ of Reading Press, Reading, pp 67–78

Tungate KD, Israel DW, Watson DM, Rufty TW (2007) Potential changes in weed competitiveness in an agroecological system with elevated temperatures. Environ Exp Bot 60:42–49

Tuong TP, Bouman BAM (2003) Rice production in water-scarce environments. In: Kijne JW, Barker R, Molden D (eds) Water productivity in agriculture: limits and opportunities for improvements. CABI Publishing, Wallingford, pp 53–67

Valerio M, Tomecek M, Lovelli S, Ziska LH (2011) Quantifying the effect of drought on carbon dioxide-induced changes in competition between a C_3 crop (tomato) and a C_4 weed (*Amaranthus retroflexus*). Weed Res 51:591–600

Valerio M, Tomecek M, Lovelli S, Ziska LH (2013) Assessing the impact of increasing carbon dioxide and temperature on crop-weed interactions for tomato and a C_3 and C_4 weed species. Eur J Agron 50:60–65

Varanasi A, Prasad PVV, Jugulam M (2015) Impact of climate change factors on weeds and herbicide efficacy. Adv Agron 135:107–146

Vengris J, Colby WG, Drake M (1955) Plant nutrient competition between weeds and corn. Agron J 47:213–216

Viger PR, Eberlein CV, Fuerst EP, Gronwald JW (1991) Effects of CGA-154281 and temperature on metolachlor absorption and metabolism, glutathione content, and glutathione-s-transferase activity in corn. Weed Sci 39:324–328

Walter KJ, Armstrong KV (2014) Benefits, threats and potential of *Prosopis* in South India. Forests Trees Livelihoods 23(4):232–247

Wand SJE, Midgley GF, Jones MH, Curtis PS (1999) Responses of wild C_4 and C_3 grass (Poaceae) species to elevated atmospheric CO_2 concentration: a meta-analytic test of current theories and perceptions. Glob Chang Biol 5:723–741

Wang M, Wu Y, Elgowainy A (2007) Operating manual for GREET: version 1.7. Center for Transportational Research, Energy Systems Division, Argonne National Laboratory, Argonne, 149

Ward JK, Tissue DT, Thomas RB, Strain BR (1999) Comparative responses of model C_3 and C_4 plants to drought in low and elevated CO_2. Glob Chang Biol 5:857–867

Wayne P, Foster S, Connolly J, Bazzaz F, Epstein P (2002) Production of allergenic pollen by ragweed (*Ambrosia artemisiifolia* L.) is increased in CO_2-enriched atmospheres. Ann Allergy Asthma Immunol 8:279–282

Weart SR (2003) The discovery of global warming. Harvard University Press, Cambridge, MA, 228pp

Webb AAR, McAinsh MR, Mansfield TA, Hetherington AM (1996) Carbon dioxide induces increases in guard cell cytosolic free calcium. Plant J 9:297–304

Wheeler TR, Craufurd PQ, Ellis RH, Porter JR, Vara Prasad PV (2005) Temperature variability and the yield of annual crops. Agric Ecosyst Environ 82:159–167

Wiese AF, Vandiver CW (1970) Soil moisture effects on competitive ability of weeds. Weed Sci 18:518–519

Willingham SD, Senseman SA, McCauley GN, Chandler JM (2008) Effect of temperature and propanil on penoxsulam efficacy, absorption, and translocation in alligatorweed (*Alternanthera philoxeroides*). Weed Sci 56:780–784

Woodward FI, Williams BG (1987) Climate and plant distribution at global and local scales. Plant Ecol 69(1):189–197

Yoshie F (1986) Intercellular CO_2 concentration and water use efficiency of temperate plants with different life forms and from different microhabitats. Oecologia 68:370–374

Zhou J, Tao B, Messersmith CG, Nalewaja JD (2007) Glyphosate efficacy on velvetleaf (*Abutilon theophrasti*) is affected by stress. Weed Sci 55:240–244

Zhu C, Zeng ZLH, Zhu J, Xie Z, Liu G (2008) Effect of nitrogen supply on carbon dioxide-induced changes in competition between rice and barnyard grass (*Echinochloa crus-galli*). Weed Sci 56:66–71

Ziska LH (2000) The impact of elevated CO_2 on yield loss from a C_3 and C_4 weed in field-grown soybean. Glob Chang Biol 6:899–905

Ziska LH (2001) Changes in competitive ability between a C_4 crop and a C_3 weed with elevated carbon dioxide. Weed Sci 49:622–627

Ziska LH (2003a) Evaluation of the growth response of six invasive species to past, present and future atmospheric carbon dioxide. J Exp Bot 54(381):395–404

Ziska LH (2003b) Climate change, plant biology and public health. World Res Rev 15:271–288

Ziska LH (2003c) Evaluation of yield loss in field sorghum from a C_3 and C_4 weed with increasing CO_2. Weed Sci 51:914–918

Ziska LH (2004) Rising carbon dioxide and weed ecology. In: Inderjit (ed) Weed biology and management. Kluwer Academic, Dordrecht, pp 159–176

Ziska LH (2010) Elevated carbon dioxide alters chemical management of Canada thistle in no-till soybean. Field Crop Res 119:299–303

Ziska LH (2014) Increasing minimum daily temperatures are associated with enhanced pesticide use in cultivated soybean along a latitudinal gradient in the mid-western United States. PLoS One 9(6):e98516

Ziska LH (2016) The role of climate change and increasing atmospheric carbon dioxide on weed management: herbicide efficacy. Agric Ecosyst Environ 231:304–309

Ziska LH, Bunce JA (1995) Growth and photosynthetic response of three soybean cultivars to simultaneous increases in growth temperature and CO_2. Physiol Plant 94:575–584

Ziska LH, Bunce JA (1997) Influence of increasing carbon dioxide concentration on the photosynthetic and growth stimulation of selected C_4 crops and weeds. Photosynth Res 54:199–208

Ziska LH, Bunce JA (2006) Plant responses to rising carbon dioxide, In: Plant Growth and Climate Change (Eds: Morison JIL, Morecroft MD), Oxford-Blackwell Publishing Ltd. pp 17–47

Ziska LH, Bunce JA (2007) Predicting the impact of changing CO2 on crop yields: Some thoughts on food. New Phytol 175: 607–618

Ziska LH, Caulfield FA (2000) Rising CO_2 and pollen production of common ragweed (*Ambrosia artemisiifolia*), a known allergy-inducing species: implications for public health. Aust J Plant Physiol 27:893–898

Ziska LH, Dukes J (2011) Weed management: herbicides. In: Weed biology and climate change. Wiley-Blackwell, Ames, pp 143–162

Ziska LH, Goins EW (2006) Elevated atmospheric carbon dioxide and weed populations in glyphosate treated soybean. Crop Sci 46:1354–1359

Ziska LH, McClung A (2008) Differential response of cultivated and weedy (red) rice to recent and projected increases in atmospheric carbon dioxide. Agron J 100:1259–1263

Ziska LH, McConnellLL (2016) Climate change, carbon dioxide, and pest biology: monitor, mitigate, manage. J Agric Food Chem 64:6–12

Ziska LH, Runion GB (2007) Future weed, pest and disease problems for plants. In: Newton PCD, Carran RA, Edwards GR, Niklaus PA (eds) Agroecosystems in a changing climate. CRC Press, Boca Raton, pp 261–287

Ziska LH, Teasdale JR (2000) Sustained growth and increased tolerance to glyphosate observed in a C_3 perennial weed, quackgrass (*Elytrigia repens*), grown at elevated carbon dioxide. Aust J Plant Physiol 27:159–166

Ziska LH, Weerakoon W, Namuco OS, Pamplona R (1996) The influence of nitrogen on the elevated CO_2 response in field grown rice. Aust J Plant Physiol 23:45–52

Ziska LH, Teasdale JR, Bunce JA (1999) Future atmospheric carbon dioxide concentrations may increase tolerance to glyphosate. Weed Sci 47:608–615

Ziska LH, Faulkner SS, Lydon J (2004) Changes in biomass and root:shoot ratio of field-grown Canada thistle (*Cirsium arvense*), noxious, invasive weed, with elevated CO_2: implications for control with glyphosate. Weed Sci 52:584–588

Ziska LH, Reeves JB III, Blank B (2005) The impact of recent increases in atmospheric CO_2 on biomass production and vegetative retention of cheatgrass (*Bromus tectorum*): implications for fire disturbance. Glob Chang Biol 11:1325–1332

Ziska LH, Sicher RC, George K, Mohan JE (2007) Rising atmospheric carbon dioxide and potential impacts on the growth and toxicity of poison ivy (*Toxicodendron radicans*). Weed Sci 55:288–292

Ziska LH, Epstein PR, Schlesinger WH (2009) Rising CO_2, climate change, and public health: exploring the links to plant biology. Environ Health Perspect 117:155–158

Ziska LH, Tomecek MB, Gealy DR (2010) Competitive interactions between cultivated and red rice as a function of recent and projected increases in atmospheric carbon dioxide. Agron J 102:118–123

Ziska LH, Blumenthal DM, Runion GB, Hunt ER Jr, Diaz-Soltero H (2011) Invasive species and climate change: an agronomic perspective. Clim Change 105:13–42

Ziska LH, Gealy DR, Tomecek MB, Jackson AK, Black HL (2012) Recent and projected increases in atmospheric CO_2 concentration can enhance gene flow between wild and genetically altered rice (*Oryza sativa*). PLoS One 7(5):e37522

Climate Change and Weeds of Cropping Systems

Annie Ruttledge and Bhagirath S. Chauhan

Introduction

The impact of weeds is diverse and multifaceted, the consequences of which are experienced from across all levels of society. Weeds impact our society in the form of direct expenditure (i.e. financial resources devoted to biosecurity, quarantine and control measures), reduced economic output (i.e. lower yields, livestock losses, reduced quality of farm products, forgone production, etc.) and welfare costs (i.e. impacts on human health, damage to natural assets and biodiversity). Economic losses due to introduced weeds in cropping and pasture systems in Australia, the United States (US), Brazil, South Africa, India and the United Kingdom (UK) have been estimated at US$ 94.92 billion per year and their environmental costs at US$ 178 million (Pimentel et al. 2001). However, because there is no data available for many invasive weed species on damages and control costs (Jardine and Sanchirico 2018), and welfare costs are often omitted from economic estimates, the true societal cost of weed invasion remains unquantified. Weeds in cropping systems, which are the focus of this chapter, represent a major challenge to global food production. In developing economies, where farmers typically employ traditional weed control practices such as animal-powered cultivation and hand-weeding, crop losses due to weeds can be high. In India, estimated yield losses incurred due to weeds amounted to 35.8% in peanut (*Arachis hypogaea* L.), 31.4% in soybean (*Glycine max* (L.) Merr.), 30.8% in green gram (*Vigna radiata* (L.) R. Wilczek, 27.6% in pearl millet (*Pennisetum glaucum* (L.) R. Br., 25.3% in maize (*Zea mays* L.), 25.1% in sorghum

A. Ruttledge
Department of Agriculture and Fisheries, Toowoomba, QLD, Australia
e-mail: Annemieke.Ruttledge@daf.qld.gov.au

B. S. Chauhan (✉)
The Centre for Crop Science, Queensland Alliance for Agriculture and Food Innovation (QAAFI), The University of Queensland, Gatton, QLD, Australia
e-mail: b.chauhan@uq.edu.au

K. Jabran et al. (eds.), *Crop Protection Under Changing Climate*,
https://doi.org/10.1007/978-3-030-46111-9_3

(*Sorghum bicolor* (L.) Moench), 23.7% in sesame (*Sesamum indicum* L.), 21.4% in mustard (*Brassica nigra* L.), 21.4% in direct-seeded rice (*Oryza sativa* L.), 18.6% in wheat (*Triticum aestivum* L.) and 13.8% in transplanted rice in experiments where weeds were controlled according to standard farming practice (Gharde et al. 2018).

In more advanced economies where modern weed control methods are applied, weeds still impose heavy yield penalties and control costs. In US agriculture, weeds cause a 12% reduction in crop yields, equivalent to US$ 33 billion in lost crop production annually based on the potential value of all US crops being in excess of US$ 267 billion/year (Pimentel et al. 2005). In Australian grain cropping systems, average cost on weed control was estimated to be AU$ 113/ha. Weeds cost Australian grain growers AU$ 146/ha in expenditure and losses, with this annual loss in yield amounting to 2.76 million tonnes of grain (Llewellyn et al. 2016).

These estimates clearly demonstrate that weeds reduce the profitability of cropping systems, even in situations where weeds are subjected to high levels of control. In the absence of control measures, weeds have greater potential to reduce global crop production than any other type of crop pest (Oerke 2006). For example, soybean is grown over 32×10^6 hectares in North America and valued in excess of US$ 33 billion, but soybean production and economic value would be halved if weeds were left uncontrolled (Soltani et al. 2017). Similarly, uncontrolled weed presence in maize across the United States and Canada has potential to cause 50% yield loss, equating to an amount of 148 million tonnes of maize worth over US$ 26.7 billion annually (Soltani et al. 2016). For dry bean (*Phaseolus vulgaris* L.) production in North America, decrease in yield because of uncontrolled weeds resulted in an average yield loss of 71.4%: a reduction of 941×10^6 and 184×10^6 kg for the United States and Canada, valued at approximately US$ 622 and US$ 100 million, respectively (Soltani et al. 2018). These estimates clearly illustrate the importance of continued research and extension of support for effective weed management in the future.

In this chapter, we review the relevant literature with the objective of summarising the key ways in which weeds and crop species are predicted to respond to climate change in the context of cropping systems. We then outline the likely implications for crop-weed competitive interactions and discuss how climate change could influence which weed control technologies will remain viable and effective in the future.

Climate Change and Cropping Systems

Temperature changes are among the foremost concerns associated with climate change. The Earth is experiencing rising temperatures due to the combustion of fossil fuels which emit carbon dioxide (CO_2) (Wijffels et al. 2016). Anthropogenic emissions of non-CO_2 greenhouse gases, such as nitrous oxide and methane, also contribute considerably to warming (Montzka et al. 2011).

Temperature is a primary factor affecting plant development. Reproductive processes and pollination, in particular, are sensitive to temperature extremes across all species of crops, greatly affecting production systems (Hatfield and Prueger 2015). Warmer temperatures are already causing yield loss in major crops (Chen et al. 2016; Gourdji et al. 2013), and in the absence of adaptation on an unprecedented scale, extreme heat will have a significant effect on productivity of agriculture (Burke and Emerick 2016).

Additionally, increased heat leads to greater evaporation, hence increasing the duration and intensity of drought. From 1964 to 2007, production losses due to drought were associated with a reduction in both harvest area and yield, whereas extreme heat mainly decreased cereal yields. Furthermore, recent droughts have been the most damaging, with developed countries suffering greater production losses than developing ones (Lesk et al. 2016). Globally, by 2100, the disaster area affected by drought will increase from 15 to 44%. The average cropland drought risk index will double from 53 to 105 under 2050 projections, increasing to 129 by 2100. Correspondingly, future climate change (related to drought disaster) will significantly increase the yield reduction rates for major crops: by >50% and almost 90% in 2050 and 2100, respectively (Li et al. 2009).

In addition to these challenges, the occurrence of warmer air with greater water holding capacity is producing more intense precipitation events. Increased water vapour in the atmosphere has resulted in the increase of extratropical rain, thunderstorms or snow storms and tropical cyclones of greater severity. Even where total precipitation is decreasing, such events are observed to be widely occurring, rendering true the saying that 'it never rains but it pours!' (Trenberth 2011). Recent flooding events have greatly damaged crop production worldwide. US production losses in maize caused by frequent occurrence of extreme weather events may become twice during the next 30 years, causing additional damages estimated at $3 billion per year (Rosenzweig et al. 2002).

Elevated atmospheric CO_2 may have some positive effects to the otherwise negative impact of rising temperature and reduced soil moisture (Lobell and Gourdji 2012). Higher CO_2 has been shown to have a fertilisation effect in C_3 species such as wheat and rice. Biomass increase has been forecasted to augment marketable yield by 8–70% for C_3 cereals and by 20–144% for cash and vegetable crops (Korres et al. 2016). However, there is concern that lower protein concentrations in harvestable yield associated with CO_2 fertilisation may reduce the nutritional quality of crops (Taub et al. 2008).

Responses of Crop Weeds to Climate Change

The large body of scientific evidence in support of climate change leaves little doubt that global temperature, precipitation and CO_2 levels are already considerably different from pre-industrial levels. Considering the limited response of humans to mitigate against the causes of climate change, there is every reason to expect current trends to continue. Modern plant taxa have been persisting for the last 2.5 million years of climate variability (Davis and Shaw 2001). Whether plant species have the capacity to respond fast enough to keep pace with the current rapid pace of climate change remains to be seen. Generally, plant species have three options to avoid extinction in a changing environment: migrate, acclimate or adapt (Lavorel and Garnier 2002; Pautasso et al. 2010; Peters et al. 2014). The discussion below will consider each of these responses in terms of weeds in cropping systems, based on the available literature.

Migration

The term migration is used here to describe the landscape-scale expansion of plant species. Weeds that migrate into new areas are commonly referred to as invasive or alien plant species. For certain invasive species, climate change is anticipated to increase the area occupied (Bradley et al. 2010; Kumaratilake and Preston 2005; McFadyen and Skarratt 1996). However, empirical evidence for landscape-scale range expansion due to climatic factors alone is scarce for arable weed species. The paucity of evidence may be due in part to the difficulty of separating any direct climate effect from changes driven by human-mediated processes. Indeed, compared with the impact of human activities in facilitating the movement of species into new areas, range expansion due to climate factors alone is likely to be minor. Moreover, in addition to crossing geographical barriers and entering a new area, plant species must also become established in a new location, a process which may take many years to become evident. Similarly, where an increase in range is observed, it is often very likely to be the consequence of natural expansion of populations, rather than a direct result of climate change (Hulme 2017). Practitioner management strategies will largely dictate the extent to which weed migration will impact arable systems. Climate change will no doubt require adaptation of cropping systems, including changed rotations, new crops and different management practices (e.g. irrigation, nutrient application, etc.), and these changes will have direct consequences for weeds, both new and naturalised species.

Acclimation

Elevated CO_2 stimulates photosynthesis and growth in C_3 weeds and reduces stomatal aperture, resulting in increased water use efficiency in both C_3 and C_4 weeds (Patterson 1995). Moreover, weed acclimation responses may render them more or less susceptible to herbicides, depending on characteristics of both the herbicides and weed species. This will be explored in more detail later in this chapter.

The ability of an introduced plant species to acclimate to an environment is influenced by phenotypic plasticity, increasing both the invasive risk and the ability to influence which weeds will come to dominate a cropping system (Scheiner 2016). Phenotypic plasticity is the ability of a population to express different phenotypic traits under varying environmental conditions (Bradshaw 1965). Richards et al. (2006) described three basic strategies that can assist plant species in becoming successful invaders. The first is a 'jack-of-all-trades' strategy, which promotes success under conditions that would generally be considered unfavourable for plant growth. Examples of crop weeds with a jack-of-all-trades genotype include knapweed (*Centaurea diffusa*) (Turner et al. 2015) and Johnsongrass (*Sorghum halepense*) (Atwater et al. 2016). A 'master-of-some' succeeds only under favourable environmental conditions. The 'master-of-some' category is well-illustrated in the context of arable fields. For example, in a study *Agrostemma githago* and *Bupleurum rotundifolium* were only found in winter cereal fields in arable lands of Western Europe (Fried et al. 2010). The third category mentioned by Richards et al. (2006) is the 'jack-and-master' strategy which, by tying the other two strategies together, can have a non-plastic fitness yet is flexible enough to increase its fitness by utilising favourable environments and so succeeds under all conditions. In this study, introduced *Phragmites* outperformed the native North American conspecific by a factor of 2–3 under every global change scenario (Mozdzer and Megonigal 2012).

Evolution

Many species of weeds are capable of relatively rapid genetic change. A classic illustration of this is the evolution of resistance to herbicides in cropping systems, including alarming cases in which weed species have developed multiple resistance to herbicides with different modes of action (Chhokar and Sharma 2008; Owen et al. 2007; Yu et al. 2007). Such capacity for rapid evolution enhances the invasive potential of a weed as well as its responsiveness to anthropogenic ecosystem modification. It has been speculated that invasive species experience a lag phase, which could be due to the time needed for them to evolve to fit the new habitat. This implies that many invasive species or their biotypes during this climate change period may be in the process of evolving adaptations that could lead to a high population growth in the near future (Clements and Ditommaso 2011).

Consequences of Climate Change for Weed Control in Arable Ecosystems

Competition Between Weeds and Crop Plants

As we have already discussed, crop productivity is already being directly impacted by climate change. The increase of environmental stress on crops reduces their competitive ability against weeds while also exposing them to increased vulnerability to attack from insects and plant pathogens (Patterson 1995). Climate change is likely to cause an extensive shift in patterns of photosynthetic limitation in higher plants. A lot of research and discussion has centred on the relative importance of C_3 and C_4 photosynthetic pathways in determining the competitive advantage of crop-weed interactions under climate change. Higher CO_2 has a fertilisation effect in C_3 species such as wheat and rice, given that photo respiratory costs in the C_3 photosynthesis pathway are eased by higher levels of CO_2 (Lobell and Gourdji 2012).

Because most of the world's major crops are C_3 species, while many major agricultural weeds are C_4 species, it is plausible that an increase in atmospheric CO_2 could alter the balance of competition in favour of crops. C_4 photosynthesis enhances carbon gain in hot environments; however, certain characters unique to C_4 photosynthesis may reduce phenotypic plasticity possibility compared with C_3 photosynthesis (Sage and McKown 2005). It is difficult to clearly ascertain whether a C_3 or C_4 photosynthetic pathway is advantageous under climate change due to the interaction of climatic factors among themselves. Alberto et al. (1996) reported that elevated CO_2 alone could increase competitiveness in a C_3 crop (e.g. rice) relative to a C_4 weed (e.g. *Echinochloa glabrescens*); the C_4 weed could be favoured with the simultaneous increases in CO_2 and temperature. Similarly, Valerio et al. (2011) observed reduced weed competition when tomato (C_3) and redroot pigweed (*Amaranthus retroflexus*, C_4) were grown under well-watered conditions, but redroot pigweed performed better when drought and high CO_2 occurred synchronously. In contrast, in a study comparing three C_4 grass weeds (*Digitaria ciliaris*, *Echinochloa crus-galli* and *Eleusine indica*) and soybean (C_3 crop), elevated CO_2 increased the growth of both C_3 and C_4 plants under drought conditions by improving their water economy, but CO_2 enrichment growth stimulation was greater in the C_3 crop (Patterson 1986).

Weed-crop competition will be influenced by interactions between climatic variables and other factors important for plant growth. In rice (C_3), biomass increased with elevated CO_2 relative to *E. crus-galli* (C_4), but only in the presence of adequate soil nitrogen (Zhu et al. 2008). In the nitrogen-limited condition, the competitive ability of rice relative to *E. crus-galli* was reduced under enriched CO_2. There are also situations where the crop and co-occurring weed species have the same photosynthetic pathway. Research comparing maize and *S. halepense*, both C_4, reported that maize had higher competitive ability when soil water was at field capacity (FC), but reducing soil water content to 75% FC reduced competitiveness of maize, and *S. halepense* populations collected from subhumid regions became more competitive under drought conditions (Leguizamón et al. 2011). The suggestion here is that genetic or phenotypic differences between weed populations could be more important than photosynthetic pathway in determining the extent of crop-weed competition.

Weed Invasion

Climate change will have a significant effect upon the processes that regulate weed invasion, in particular growth and reproduction (Hulme 2017). For example, longer periods of drought and elevated CO_2 are likely to significantly increase the growth and seed production of *Parthenium hysterophorus* and may ease the further invasion of this damaging weed in tropical and subtropical agro-ecosystems of Australia (Nguyen et al. 2017), assuming the existence of suitable dispersal pathways.

At this stage, it is relevant to point out that agricultural weeds evolve in response to crop management practices, and considerations such as crop type, agronomy and weed control exert strong selection pressures that may in some cases outweigh changes in climatic factors that could otherwise support weed invasion. Weed populations have demonstrated a capacity to evolve rapidly in response to intense, human-derived selection pressures (e.g. herbicide resistance), and it is reasonable to assert that such evolution could occur simultaneous to climate-driven adaptation. This highlights a need for studies on evolutionary biology so that weed adaptation and their spread could be minimised (Neve et al. 2009).

Herbicide Efficacy

In modern agriculture, herbicides play a crucial role in farm productivity and profitability, especially in systems that have adopted conservation agriculture (CA). Conservation tillage, which is central to CA, is vital to protecting soil structure and preventing erosion and is advocated as a means to reduce greenhouse gas emissions from agricultural land use (Busari et al. 2015). In conservation tillage systems, the ability to plant into unploughed soil, leaving crop residues on the surface, is enabled by herbicides that provide an alternative to cultivation for controlling weeds. The efficacy of many herbicides could be affected by climate change parameters, which could make weed management a difficult task for sustainable crop production in modern CA systems.

There is already sufficient evidence to suggest that climate change could significantly reduce herbicide efficacy due to direct effects on plant biochemistry and morphology, as well as field access and spray coverage being impacted by climatic extremes (Ziska 2016). Environmental factors such as temperature, soil moisture and CO_2 affect the uptake, translocation and efficacy of different herbicides. Furthermore, interactions among these environmental factors may have unpredictable effects on herbicide efficacy (Varanasi et al. 2016).

Existing literature (Table 1) on commonly used herbicides suggests that the impacts of climate change on herbicide resistance not only will vary according to the mode and type of herbicide but will also differ between species and within biotypes of the same species (i.e. resistant versus susceptible weed biotypes).

Table 1 Impact of climate change on herbicide efficacy (HE)

Climate change element	Herbicides	Weed species	Result	References
Elevated CO_2	Glyphosate and fluroxypyr-meptyl	14 species (grasses, herbs, shrubs and vines)	Species-specific	Waryszak et al. (2018).
	Glyphosate and metsulfuron methyl	15 species (grasses, herbs, shrubs and trees)	No clear relationships between species' responses to herbicide	Waryszak et al. (2018)
	Glyphosate and halosulfuron	*Cyperus rotundus* L. and *C. esculentus* L.	Nil reduction in HE	Marble et al. (2015)
	Glyphosate	Four C_4 invasive exotic grasses	Higher tolerance in 3 out of 4 grasses when mature, due to increased biomass production resulting in a dilution effect	Manea et al. (2011)
	Glyphosate	*Cirsium arvense*	Weed species more competitive with soybean crop and harder to control chemically in response to elevated CO_2	Ziska (2010).
	Glyphosate	*Cirsium arvense*	Reduction in glyphosate efficacy in the enriched CO_2 treatment, attributed to a dilution effect	Ziska et al. (2004).
	Glyphosate	C_3 perennial (*Elytrigia repens*)	Sustained growth and increased tolerance to glyphosate at elevated CO_2 in intermediate and older plants	Ziska and Teasdale (2000).
Combination of increased temperature and elevated CO_2	Cyhalofop-butyl	Multiple-resistant (MR) *Echinochloa colona*	HE not reduced in susceptible plants. High CO_2 and high temperature increased the resistance level of MR *E. colona*	Refatti et al. (2019)
	Glyphosate	*Bromus tectorum* L., *Hordeum murinum* L. and *Lactuca serriola* L.	Nil impact on HE	Jabran and Doğan (2018).
Temperature	Glyphosate and dicamba	*Kochia scoparia*	Decreased dicamba translocation and reduced glyphosate absorption contribute to poor control at high temperature	Ou et al. (2018)

(continued)

Table 1 (continued)

Climate change element	Herbicides	Weed species	Result	References
	Glyphosate and 2,4-D	*Ambrosia artemisiifolia* and *Ambrosia trifida*	HE of glyphosate and 2,4-D for control of both species can be improved if applied at warm temperatures (29/17 °C day/night) due to increased absorption and/or translocation compared with applications during cooler temperatures (20/11 °C day/night), regardless of susceptibility or resistance to glyphosate	Ganie et al. (2017).
	Saflufenacil	Glyphosate-resistant (GR) and glyphosate-paraquat resistant *Conyza bonariensis*	Saflufenacil- and glyphosate-alone provided better control at 15/10 °C and 25/20 °C than under 35/30 °C. Under 35/30 °C, a tank mixture of saflufenacil + glyphosate provided the best control	Dennis et al. (2016)
	Herbicides that inhibit acetyl-CoA carboxylase (ACCase)	*Alopecurus myosuroides*, *Avena sterilis*, *Lolium rigidum*, *L. multiflorum* and *Setaria viridis*	Temperature-dependent herbicide sensitivity based on metabolic detoxification, suggesting higher risks for the evolution of herbicide-resistant weeds under predicted climatic conditions	Matzrafi et al. (2016).
	Glyphosate	*Echinochloa colona*	Reduced glyphosate efficacy at high temperatures on resistant *E. colona* biotypes, making them harder to control in summer	Nguyen et al. (2016)
	Mesotrione	*Amaranthus palmeri*	Sensitivity drastically increased when temperature decreased from 32.5/22.5 to 25/15 °C day/night. Sensitivity was reduced when grown under 40/30 °C day/night compared to 32.5/22.5 and 25/15 °C	Godar et al. (2015)

(continued)

Table 1 (continued)

Climate change element	Herbicides	Weed species	Result	References
	Glufosinate	*Raphanus raphanistrum*	Low temperature reduces glufosinate activity and translocation. Control efficacy higher under warmer conditions	Kumaratilake and Preston (2005).
	Pyrithiobac	*Amaranthus palmeri*	Pyrithiobac efficacy was highest for the application within the thermal application range 20–34 °C and significantly reduced at temperatures above and below	Mahan et al. (2004)
	Isoproturon	–	Based on modelling using real weather data, accelerated decline in herbicide residues was attributed to higher soil temperature and would result in shorter duration of weed control	Bailey (2004)
Temperature and soil moisture (SM)	MON 37500	*Bromus tectorum*, *Avena fatua* and *Aegilops cylindrica*	Cool air temperatures decreased the herbicide metabolism in all species, whereas SM had no effect	Olson et al. (2000)
Temperature and RH (RH)	Mesotrione (foliar applied)	*Xanthium strumarium L.*, *Amaranthus rudis Sauer*, *Digitaria sanguinalis L. Scop.*, *Ipomoea hederacea (L.) Jacq.*, *Abutilon theophrasti (L.) Medic.*	Changes in temperature and RH can significantly alter the efficacy of mesotrione in foliar application, but the degree and nature of the effect is species dependent	Johnson and Young (2002)
	Glyphosate	*Desmodium tortuosum*	Uptake and translocation were reduced at high temperature or low RH	Sharma and Singh (2001).
	Glufosinate-ammonium	*Setaria viridis (L.) Beauv.*	HE was significantly decreased by low RH	Anderson et al. (1993)

(continued)

Table 1 (continued)

Climate change element	Herbicides	Weed species	Result	References
Soil moisture (SM)	11 pre-emergent herbicides for sunflower (*Helianthus annuus*)	*Chenopodium album*, *E. crus-galli*, *A. retroflexus*, *Mercurialis annua* and *Solanum physalifolium*	The efficacy of linuron, pethoxamid and prosulfocarb was strongly affected by SM and was reduced under dry conditions. For flurochloridone, phytotoxicity increased due to irrigation after herbicide application	Jursik et al. (2015)
	Bispyribac	*E. crus-galli*	Increasing SM conditions resulted in greater efficacy	Koger et al. (2007)
	Imazethapyr	*E. crus-galli (L.) Beauv.*, *Oryza sativa L.*	High SM conditions reduced efficacy of imazethapyr for control of *E. crus-galli* and *O. sativa* when applied as preplant incorporated. POST activity unaffected by SM conditions	Zhang et al. (2001).
	Fenoxaprop, fluazifop-P, haloxyfop, sethoxydim	*Setaria viridis*	Low soil moisture for 10 to 14 days before and 7 days after herbicide application reduced weed control with fenoxaprop, fluazifop-P, haloxyfop and sethoxydim. The activity of sethoxydim was reduced less by SM than the other herbicides. If plots were irrigated at the time of application, normal rates of these herbicides were effective	Boydston (1990)
Relative humidity (RH) and soil moisture (SM)	Fluroxypyr	*Kochia scoparia (L.) Schrad*, *Amaranthus palmeri (S.) Wats.*	*Kochia* control was not affected by RH, but control was greater when plants were grown in moist soil than in dry soil. Conversely, *A. palmeri* control was greater when plants were grown at 90% RH than at 35% RH, but control did not differ between moist and dry soils	Lubbers et al. (2007)

Of particular concern is the potential effect of climate change on glyphosate efficacy. Glyphosate-resistant (GR) weeds are already a major problem in field crops and present a significant threat to future global food production. There are concerns that climate change, particularly elevated CO_2 concentrations, could enhance glyphosate resistance mechanisms. A better knowledge of the biochemical, molecular and genetic processes by which weeds evolve resistance to glyphosate and how CO_2 affects these processes will be essential to maintaining effective weed control using glyphosate (Fernando et al. 2016). Further, there is need to investigate the effects of various climatic changes on commonly used herbicides and for all major weed species (especially resistant biotypes) (Varanasi et al. 2016).

Looking Forward: Crop Weed Management in a Changing Climate

Competitive Cultivars

Damages incurred under climate change could be more costly for agricultural production than any other human activity. Yield losses will occur due to the direct effects of climate change on crops, as well as the resulting indirect effects such as more costly and difficult to control weeds. Because competitive crops are better at sequestering light, water and other resources that are otherwise available to augment weed growth, the use of crop cultivars adapted to conditions of high climatic variability and extremes could be a solution to both climate change and pest challenges (Korres et al. 2016).

The mechanism of plant breeding presents a vital opportunity for cropping system adaptations under climate change. Atlin et al. (2017) reviewed this topic, noting that breeding for climate change has been focused on the phenology and stress tolerance of plants according to genes with large effects on heat and drought tolerance, both of which are highly polygenic. They suggested that adaptation will occur incrementally via the development of cultivars adjusted for allele frequencies at many loci through rapid-cycle breeding.

Weed-competitive cultivars have gained interest due to the challenge of managing herbicide-resistant weeds. Interest in weed competitive cultivars has been increased with the increasing challenge of managing herbicide-resistant populations. They offer the opportunity for a cost-effective component of an integrated weed management (IWM) system. Such cultivars are a recent priority for breeders and for farmer selection despite the identification of those with high competitive potential in cereal crops. Crop traits that assist in weed suppression include plant height, high biomass, speed of development (fast germination, rapid growth and canopy closure), root-shoot characteristics, canopy architecture (large leaf area), self-supportive allelopathy and partitioning of resources (Andrew et al. 2015; Dass et al. 2017; Sardana et al. 2017). Competitive crop cultivars present a tool for both negating climate change and contributing to IWM. Realising the complementarity of both areas of potential should be prioritised in breeding programmes in order to either serve both purposes or serve one without heavily penalising the other. This will most likely occur through the capture of the combined effect of multiple traits rather than that of a single trait (Andrew et al. 2015). These criteria, once identified, could be transformed into a protocol to equip researchers and agronomists to select the strongest candidates from among new and existing cultivars, to test under local conditions and against key weed species. The development of competitive cultivars will be greatly influenced by local environmental conditions and the specific challenges presented by weeds there. Such local screening should use varieties developed in the last 10 years, as older varieties were developed under a climate that is different from today, potentially placing farmers at risk (Atlin et al. 2017). By combining rapid breeding and varietal replacement with research targeted towards local regional weed challenges, farmers would have the benefit of knowing which of the available cultivars are most likely to sustain production in their particular environment, against their major weed flora.

Diverse Cropping Systems

Crop rotation is an important tool for achieving diversity in crop production. In a rotation, weeds are affected by the diversity of plant resource demands, alterations to crop management timing and allelopathy. Rotations should include crops sown in different seasons, of annual and perennial varieties, managed under different herbicides and should come from various crop families (Nichols et al. 2015). A number of crop species (e.g. rye (*Secale cereale* L.), sorghum, rice, sunflower, rape seed (*Brassica napus* L.) and wheat) are known to produce allelochemicals that have potential for weed suppression. Rotating a routine crop with an allelopathic crop cultivar for one season is one way to increase diversity in weed management and potentially reduce herbicide use. Further, several types of allelopathic plants can be used for intercropping to smother weeds (Jabran et al. 2015). At present, monocropping is the prevalent form of cropping system, but the stability of monocropping systems is being degraded by climate change. Intercropping (IC) has been shown to have higher yield stability and higher yield levels for cereals and grain legumes compared to growing the component crops in monoculture.

Cover crops, including those with allelopathic potential against weeds, provide multiple benefits to cropping rotations though soil, agricultural production and environmental factors. Studies in temperate soils by Blanco-Canqui et al. (2015) found that cover crops increased soil organic C stocks, reduced runoff and sediment loss, diminished wind erosion potential, alleviated soil compaction, improved soil structural and hydraulic properties, moderated soil temperature, improved microbial properties, recycled nutrients and suppressed weeds. However, the magnitude of such benefits varies according to location. Whether there is any benefit to crop yield is still debatable, and cover cropping can reduce yields in water-limited regions by reducing available water for subsequent crops. Adoption of cover cropping has been slow among conventional growers. Cover crops are becoming an important tactic for managing GR *A. palmeri* in cotton and soybean in the southern United States, though its use as a single, season-long tactic is not recommended (Wiggins et al. 2015). Further integration of cover crops into cropping systems will require ongoing research to evaluate their multi-functionality for different climates and management scenarios, as well as the relevant short- and long-term economic returns associated. The development of germplasm with better characteristics, such as cover crops that self-destruct after suppressing weeds and prior to competing with crops, presents one area of potential in plant breeding (Shaner and Beckie 2014).

Manipulating Crop-Weed Competition

Seeding rate, row spacing and row direction are underutilised yet impactful, environmentally sustainable weed control methods (Sardana et al. 2017). In aerobic rice, narrow row spacing (Chauhan and Johnson 2011) and increased seeding rate (Anwar et al. 2011) have shown the potential to reduce weed pressure without sacrificing yield. While not sufficient to suppress weeds completely, these practices can achieve up to a 50% reduction in herbicide use (Dass et al. 2017). Use of wide row spacing in soybean requires earlier weed management programmes due to reduced crop tolerance in the early growth phase (Knezevic et al. 2003); however, doubling the seeding rate for conventional wide row planting can improve weed suppression (Liebert and Ryan. 2017). Research in canola has demonstrated that higher seeding rates resulted in greater emergence, early crop biomass and seed weight and oil content. It also reduced the number of days to start flowering and for reaching maturity, as well as stubble density at harvest (Harker et al. 2015). Because higher seeding rates speed up flowering and reduced the number of days to maturity, this strategy could also lower the risk of exposure to high temperatures that can negatively affect flowering and pod development in canola.

In Australian winter cropping systems, where management of herbicide-resistant *Lolium rigidum* is a major challenge, the weed species was suppressed at higher wheat densities but was not affected by row spacing (Lemerle et al. 2013). Sowing wheat and barley crops in an east-west (EW) direction reduced the availability of photosynthetically active radiation (PAR) compared with a north-south (NS) direction and resulted in reduced *L. rigidum* fecundity (average of 2968 and 5705 *L. rigidum* seeds m^{-2} in the EW and NS crops). In the same study, it was found that seeding rate did not affect PAR despite the occurrence of reduced fecundity in *L. rigidum* under conditions of high seeding rate (average of 3354 and 5092 seeds m^{-2} in the crops with high and low seeding rate) (Borger et al. 2016). De Vita et al. (2017) examined the combined effects of row spacing and wheat cultivar type (tall and semi-dwarf cultivars) and found that reduced inter-row spacing resulted in less weed biomass in both the tall and semi-dwarf cultivars. While weed presence reduced grain yield in the semi-dwarf cultivars, yield and N uptake increased under narrow row spacing. In contrast, yield of the tall wheat cultivar did not vary with weed biomass or inter-row spacing (De Vita et al. 2017).

Competitive planting strategies may not be suitable in all situations. In maize, narrower row spacing does not appear to affect weed density or growth (Johnson and Hoverstad 2002) and likewise has no impact on the critical period for weed control (CPWC) or crop competitiveness with late-emerging weeds (Norsworthy and Oliveira 2004). In organic wheat and barley farming systems, increased seeding rate and the use of two-pass sowing to achieve more even crop distribution can promote greater competitiveness against weeds relative to standard production practice (Kolb et al. 2010, 2012). Additionally, the use of wide rows with inter-row hoeing was recommended due to the higher input cost of organic seed (Kolb et al. 2010, 2012).

Sowing strategies such as increased seeding rates, row spacing and row orientation recognise that crops and weeds share the same above-ground resources, competing for sunlight and space. However, weeds also compete with crops for underground (soil) resources: chiefly water and nutrients. The below-ground competition between weeds and crop plants could be managed according to fertiliser application and water management for achieving higher net returns (Kaur et al. 2018). For example, high amounts of N fertiliser, as well as the proper selection of herbicides, were found to help suppress individual weed species in soybean fields (Pinke et al. 2016). However, providing additional fertiliser and/or water to compensate for weed interference will often prove unfeasible. There is, however, benefit in understanding how soil resource pool diversity may mediate below-ground competition between weeds and crops (Smith et al. 2010). Due to technological advancements, farmers in modern farming systems have increasing levels of access to highly detailed spatial information about the soil in their fields that can be used to build an understanding of where soil resources are distributed in a cropping field (i.e. areas of high, intermediate or low nutrient/soil moisture) and how this overlaps with spatial information on yield and weed density. Applying information in this manner has potential for strategic weed management in precision agriculture, targeting weed control measures where there is the greatest cost-benefit ratio. As climate change is already resulting in reduced precipitation, higher temperatures and prolonged droughts, judicious management of soil water resources will become increasingly important.

Improved Tillage

Mechanical or physical weed control has been practiced since ancient times, and traditional practices such as hand-weeding and using animals to drag tillage implements are still common practice in many less developed parts of the world. In modern cropping systems, the importance of mechanical weed control has declined in the wake of CA. CA is a more sustainable, environmentally friendly crop management system that utilises cropping rotations, minimal soil disturbance (low-tillage or NT) and permanent crop residue soil cover (Hobbs et al. 2007). There are, however, side effects attributed to the adoption of continuous NT practices such as the build-up of herbicide-resistant weed populations (Dang et al. 2015). A return to deep tillage using a mouldboard plough is becoming increasingly commonplace in the management of major herbicide-resistant weeds, such as *Alopecurus myosuroides* Huds. in the United Kingdom, *L. rigidum* in Australia or *A. palmeri* in the southern United States. Deep tillage operations can have devastating impacts on the quality of fragile or poorly fertile soils. With many farmers returning to tillage in order to manage problems associated with continuous NT, there is a need to develop technologies for less intensive tillage before the gains in soil quality as a result of CA are lost (Shaner and Beckie 2014).

Rather than returning to routine use of tillage, the implementation of targeted tillage (strategic tillage) presents a potential tool for weed management in CA systems (Chauhan and Walsh 2017). It appears that, at least in some soils, strategic tillage can be practiced for weed control without consequences for productivity or soil health. The use of either a chisel cultivator or disc chain on a Grey Vertosol managed under NT for 15 years found that strategic tillage, irrespective of tillage timing, could be used for weed management without impacting agronomic productivity or damaging the physical, chemical or biological characteristics of the soil (Liu et al. 2016).

Further, simulation modelling has shown that even occasional strategic tillage to help control weeds in Dermosols and Sodosols may increase their susceptibility to water, sediment and nutrient losses in runoff after heavy rainfall (Melland et al. 2017). This is an important consideration given that increased frequency and severity of extreme weather events are a feature of climate change. Climate change modelling also identified strategic tillage to cause significant increases in CO_2 emissions from the Sodosol as well as decreased CH_4 absorption from both soils (Melland et al. 2017).

It should be stated that the discussion above relates to tillage as it is practiced on a paddock scale. Technologies are being developed with the potential to deploy tillage at the scale of 1 to several weed plants, using robotic technology equipped with camera sensors. Application of tillage on such small areas is less likely to have negative repercussions for soil health or plant available water and represents a valuable tool for patch management of herbicide-resistant populations.

Harvest Weed Seed Control

Originally pioneered in Australia 30 years ago, harvest weed seed control (HWSC) practices target the weed seed collected and concentrated in harvest residues. Despite low initial adoption rates, 43% of Australian grain farmers now utilise HWSC since the introduction of new weed seed targeting systems (Walsh et al. 2017). Research in North America has demonstrated the potential of HWSC for *A. palmeri* and tall waterhemp (*Amaranthus tuberculatus*) in soybean (Schwartz et al. 2016, 2017). The inclusion of HWSC tactics such as narrow-windrow burning, chaff tramlining, chaff decks, chaff carts, bale-direct and seed destruction/termination when integrated with other management strategies has been shown to diminish subsequent crop interference and is an important tactic in managing herbicide-resistant weed populations (Walsh and Powles 2014).

Because HWSC tactics target weed seeds in harvest residues, they are used to best effect on weeds that bear their seeds at harvest time and at harvest height. Based on these criteria, the seed bank of many weed species can be reduced through the use of HWSC. However, even for these species, seed retention at harvest will vary. For example, in weeds such as *Avena* spp., which can emerge in multiple cohorts over the growing season, there is a strong possibility that a proportion of plants will mature ahead of crop harvest. Sowing date and season will also impact the rate of weed seed maturity in relation to crop maturity. Furthermore, it is possible that climate change could drive modifications to cropping systems, including altered time of sowing, and that this would impact the amount of weed seed present at harvest and hence the efficacy of HWSC systems. Such considerations should be factored into weed control decisions.

Agricultural Robotics

Precision weed control is a nascent industry built upon the use of automation, mechanised vision and global position system information to identify weeds in cropping systems (Westwood et al. 2018). In certain crops, it is possible to use drones for monitoring weed populations at early growth stages or to define zones of high risk of infestation based on drone images obtained at the end of the growing season (Fernández-Quintanilla et al. 2018). Ground robots have tremendous potential for precision weed management. Agricultural equipment such as tractors and sprayers have been robotised over the last few years through the introduction of auto-guidance systems and are becoming increasingly autonomous (Vougioukas 2019). Robotic smart implements mounted to farm machinery are revolutionising herbicide application. For example, rather than continue to apply herbicide over an entire field, the French company Bilberry has developed a precision spraying system incorporating camera networks embedded directly onto spray booms together with software which can recognise and accurately spot-spray weed plants present within crops. This occurs simultaneous to the gathering of data that is automatically transcribed into maps for future processing and/or follow-up weed control operations (Bilberry 2019). Such application of spatial data lends itself to site-specific weed management in precision agriculture.

There is a general trend towards building lightweight robots to replace heavy machines. A small number have been released commercially for application within specialty crops. Many specialty crops do not have safe, effective herbicides, and producers have been heavily dependent on hired labour for hand-weeding to achieve commercially acceptable weed control.

Labour shortages and higher manual weeding costs have driven a search for other solutions. Already commercially available intrarow robotic weeders are promising tools. These tools are less regulated and more cost-effective (Fennimore and Cutulle 2019; Fennimore et al. 2016). There is also promising research on the development of robotic in-row control that can control weeds with a very limited amount of herbicides, including glyphosate and iodosulfuron (Utstumo et al. 2018). Small robots for scouting and weed control have also been released for application in broad acre situations. For example, 'SwarmFarm robots', equipped with WeedIT optical sprayer technology, have also been commercially released in Australia and have been deployed on grain farms and turf farms and in environmentally sensitive mining areas (Weedsmart 2017).

While there is still progress to be made towards the realisation of fully automatic weed control machines, the rapidity of recent advances implies that realisation of this goal may not be far off, at least in modern-precision agriculture systems. Important improvements include the development of algorithms to recognise more weed species at different growth stages and varying densities, in different crops as well as under variable environmental conditions. Agricultural environments impose challenges for robots, including the diversity and complexity of plant canopy structures (Vougioukas 2019). This last point is significant with respect to climate change, because weed species may present different characteristics through acclimation or adaptation.

Automatic weed removal technology would be improved by the additional capacity to use herbicides and non-chemical weed tactics. Currently, most available robotic weeders and advanced prototypes rely on herbicides to control weeds. Climate change has the potential to diminish weed control efficacy using herbicides, thus making robotic weeders that can selectively deploy multiple, diverse weed control tactics (e.g. cultivation, lazers, microwaves and abrasives) more responsive to climate driven limitations for weed control.

Engineers and scientists developing automation and robotic solutions to improve weed detection and control efficacy should also be guided by economic and social research. Cost-benefit analysis data associated with the use of automation and robotics on farms is needed. Such research needs to consider not only in the context of intensive agriculture in the modern cropping systems of developed countries but also for the reality of the developing world vulnerable to significant crop losses as a result of climate change (Lowenberg-DeBoer et al. 2018; Lesk et al. 2016).

Nanotechnology

Nanotechnology is emerging as a significant tool in modern agriculture. Nanotechnology presents opportunities for a targeted and an efficient application of herbicides that will make it possible to control weeds with low doses. The use of nanocoatings, biosensors and weed seed coating penetration presents different weed management opportunities in nanotechnology (Shaner and Beckie 2014).

Nanopolymer coating could put crop seeds on more equal footing with weed species, many of which are already equipped with similar abilities due to genetic diversity and plasticity.

Biological Weed Control

It is anticipated that biological weed control methods will play a greater role by 2050 thanks to their relatively low research and usage costs, their long-term efficacy and reduced environmental impact compared to several herbicides (Westwood et al. 2018). The management of regularly disturbed soils in cropping systems is generally not suitable to classical biocontrol. Bioherbicides, however, present a number of opportunities in cropping systems through increased target specificity, rapid weed degradation, seedbank management and delayed weed growth (Westwood et al. 2018). Fewer than 20 bioherbicides are currently available in the market, and the market share of bioherbicides represents less than 10% of all biopesticides (Cordeau et al. 2016); however, their share will grow in the future (Cordeau et al. 2016).

Aside from the extension-related challenges for widespread adoption of this technology, there are other challenges to the use of bioherbicides. The capacity to control multiple rather than single weed targets is not yet available, hindering the relevance and usage of microbial herbicides. Improvement in technical aspects is also required, such as product's half-life, storage requirements and the need for specialised application technology (in some cases). Perhaps the most pressing need is for the uncontrollable reaction of the biotic agent with climate conditions to be more robust in bioherbicides.

Herbicides

Present-day cropping systems are under threat due to the presence of weeds resistant to glyphosate and other herbicides. Instances of weed resistance are present in all modes of action (MOA) for current herbicides, all of which are over 20 years old. New herbicides with old MOA and the development of transgenic crops that are resistant to a wide range of existing herbicides are only short-term solutions (Westwood et al. 2018). The discovery of new herbicide MOA has long been desired, although there have been recent developments towards the commercial release of products with two new MOA. It is extremely unlikely that the discovery of novel herbicides will occur with sufficient frequency to keep pace with burgeoning herbicide resistance in major crop weeds. There is, however, cause for optimism that herbicide-resistant populations can be better managed through improved herbicide formulations (in combination with other non-chemical practices). As discussed earlier, nanotechnology can be utilised to develop better formulated herbicides.

Another emerging potential is the use of RNA interference (RNAi) technology. Such technology could eventually be used to kill weeds outright, with current emphasis on the use of RNAi to enhance susceptibility of herbicide-resistant weeds to glyphosate and other herbicides.

While it is not within the remit of this chapter to detail how herbicides should be applied for maximum efficacy, it is worth noting that in a changing climate, considerations such as the strategic timing of herbicide application will become increasingly important. Field and growth chamber experiments determined control of velvetleaf (*A. theophrasti*) to be consistently improved with glyphosate applications during light periods (Waltz et al. 2004). Field applications of flumiclorac at different times (6 am, 2 pm and 10 pm) strongly affected soybean tolerance and weed control. Herbicide application in the morning caused the greatest soybean injury (Fausey and Renner 2001). Herbicide efficacy is also affected by season, growth-stage and temperature. For example, glyphosate and saflufenacil alone provided higher efficacy on *C. bonariensis* at cooler than under warmer temperature conditions (Dennis et al. 2016).

Other Tools for Resistant Weed Management

While research into relationships between climate and weed resistance is still in the early stages, there are indications that efficacy of herbicides could be reduced in response to elevated CO_2, warmer temperatures and water scarcity. Moreover, changes in CO_2 and temperature have been shown to increase the resistance level of MR *E. colona* (Refatti et al. 2019). For this reason, any discussion of weeds and climate change needs to consider tools with potential to assist in predicting, if not overcoming, weed resistance.

Shaner and Beckie (2014) have conducted a detailed review of herbicide-resistant weed management (RWM) according to the intersection of weed control and technology. In addition to the tools and strategies already mentioned in this chapter, their review highlights the potential of 'omics' (e.g. genomics, metabolomics, etc.) to usher in a wide range of novel weed management strategies. Further, weed management programmes can benefit by predicting the long-term responses of weed species and their biotypes to changing global climate. Adding data on intrapopulation and interpopulation variability is an important first step towards assessing the potential for weed adaptation, as well as for improved accuracy of forecast which weeds are likely to present the greatest challenges in a changing climate (Neve et al. 2009).

Integrated Weed Management

The challenge of managing herbicide-resistant weeds, in combination with heightened social and regulatory pressure to reduce chemical use in food production, necessitates a move towards crop protection programmes based on integrated weed management (IWM) principles. Through integrating a suite of chemical and non-chemical weed control tactics, IWM has the potential to maintain manageable population levels, decrease the environmental severity of singular tactics, enhance the diversification and sustainability of cropping systems and lower herbicide resistance evolution risk by reducing selection pressure (Harker and O'donovan 2013).

However, implementation of IWM without herbicides has been poor, and there is little evidence of reduction in herbicide use at a global scale. Compared with herbicides, there are perceptions that non-chemical methods provide little visual evidence of success, are complex to manage and are less effective, more variable and expensive, and not able to reduce the requirement for herbicides (Moss 2010). Moreover, while there has been a notable increase in research into non-herbicidal weed management, chemical weed control remains on top (Harker and O'donovan 2013). As is the case of herbicide-resistant weed management, major obstacles remain for broad-scale implementation of IWM. The willingness and prioritisation of researchers, growers' decisions to accept recommended practices, the marketing influence of agrichemical bodies, commitment from government and a multidisciplinary approach to further research and development all impact the future trajectory of IWM (Shaner and Beckie 2014). Further, the optimal rate of engagement and adoption of IWM require a public-private partnership representing growers from all practicing countries where all essential entities are fully consulted and adequately resourced (Young et al. 2017).

Conclusion

Changes in atmospheric CO_2, global temperature and precipitation are occurring at a rate exceeding previous climate change forecasts. These climatic factors are key drivers of plant growth; and weeds, like all other plant species, will need to respond to climate change in order to survive. Weed species are by their very nature survivors, able to relocate, acclimate or adapt to changing environmental conditions. While it is yet unknown what the exact consequences of climate change will be for crop, weed interactions, it is likely that weeds, with their superior capacity for survival, will have natural competitive advantages over crop species.

The outcome for plant life in general is a matter of deep concern, as few plant species can thrive under conditions where extreme weather, including drought, will occur with greater frequency. In the future, climate change will produce an environment that is more inhospitable to crop growth than has previously occurred in Earth's history. Under such conditions, it has never been more important to provide crops with every possible advantage over weeds. The development of crop cultivars that are more resilient to climate change conditions is key for ensuring sustainable food production. These crops are also central to weed management, as healthy crops provide better competition against weeds.

While there are exciting opportunities for improved weed control through innovations such as robotics, nanotechnology and RNAi herbicide formulations, their true potential (and true costs) remains largely unknown. Weed research has already delivered a large suite of available tools that, if properly integrated, can drive down weed numbers and manage the risks of herbicide resistance. Diversity in cropping systems needs to be emphasised, incorporating the rotation of a range of species. Competitive planting strategies such as narrow row spacing, row orientation and increased seeding rate can support both crop yield and weed competitiveness. Harvest weed seed control strategies, where weed seeds mature with the crop and are captured in harvest residues, have the demonstrated capacity for managing herbicide-resistant species. Strategic tillage is another potentially valuable tool for patch management, as returning to full tillage would increase agricultural emissions of greenhouse gases. Additionally, the risks of devastating soil loss and structural decline are significant considering projections that drought and extreme weather events will occur with greater frequency and intensity. Herbicides will continue to play an important role in weed management for the foreseeable future; however, there exists a very real possibility that climate change will reduce herbicide efficacy, especially in weed species that have already evolved tolerance or resistance to herbicides.

In closing, though systems dependent on monocropping and herbicide reliance continue to dominate modern food production, increasing diversity in cropping systems presents the best defence against weeds under current and future climate scenarios. Without adoption of IWM at unprecedented rates, from both practitioners and researchers alike, the dual challenges of climate change and herbicide resistance are likely to seriously undermine the sustainability of crop production for future generations.

References

Alberto AM, Ziska LH, Cervancia CR, Manalo PA (1996) The influence of increasing carbon dioxide and temperature on competitive interactions between a C3 crop, rice (*Oryza sativa*) and a C4 weed (*Echinochloa glabrescens*). Funct Plant Biol 23(6):795–802

Anderson DM, Swanton CJ, Hall JC, Mersey BG (1993) The influence of temperature and relative humidity on the efficacy of glufosinate-ammonium. Weed Res 33(2):139–147

Andrew IKS, Storkey J, Sparkes DL (2015) A review of the potential for competitive cereal cultivars as a tool in integrated weed management. Weed Res 55(3):239–248

Anwar P, Juraimi AS, Puteh A, Selamat A, Man A, Hakim A (2011) Seeding method and rate influence on weed suppression in aerobic rice. Afr J Biotechnol 10(68):15259–15271

Atlin GN, Cairns JE, Das B (2017) Rapid breeding and varietal replacement are critical to adaptation of cropping systems in the developing world to climate change. Glob Food Sec 12:31–37

Atwater DZ, Sezen UU, Goff V, Kong W, Paterson AH, Barney JN (2016) Reconstructing changes in the genotype, phenotype, and climatic niche of an introduced species. Ecography 39(9):894–903

Bailey SW (2004) Climate change and decreasing herbicide persistence. Pest Manage Sci 60(2):158–162

Bilberry (2019). https://bilberry.io/#Features

Blanco-Canqui H, Shaver TM, Lindquist JL, Shapiro CA, Elmore RW, Francis CA, Hergert GW (2015) Cover crops and ecosystem services: insights from studies in temperate soils. Agron J 107(6):2449–2474

Borger CPD, Hashem A, Powles SB (2016) Manipulating crop row orientation and crop density to suppress *Lolium rigidum*. Weed Res 56(1):22–30

Boydston RA (1990) Soil water content affects the activity of four herbicides on green foxtail (*Setaria viridis*). Weed Sci 38(6):578–582

Bradley BA, Wilcove DS, Oppenheimer M (2010) Climate change increases risk of plant invasion in the Eastern United States. Biol Invasions 12(6):1855–1872

Bradshaw AD (1965) Evolutionary significance of phenotypic plasticity in plants. In: Advances in genetics, vol 13. Academic Press, pp 115–155

Burke M, Emerick K (2016) Adaptation to climate change: evidence from US agriculture. Am Econ J Econ Pol 8(3):106–140

Busari MA, Kukal SS, Kaur A, Bhatt R, Dulazi AA (2015) Conservation tillage impacts on soil, crop and the environment. Int Soil Water Conserv Res 3(2):119–129

Chauhan BS, Johnson DE (2011) Row spacing and weed control timing affect yield of aerobic rice. Field Crop Res 121(2):226–231

Chauhan BS, Walsh M (2017) A new approach to weed control in cropping systems. In: Biology, physiology and molecular biology of weeds. CRC Press, pp 45–62

Chen S, Chen X, Xu J (2016) Impacts of climate change on agriculture: evidence from China. J Environ Econ Manag 76:105–124

Chhokar RS, Sharma RK (2008) Multiple herbicide resistance in littleseed canarygrass (*Phalaris minor*): a threat to wheat production in India. Weed Biol Manage 8(2):112–123

Clements DR, Ditommaso A (2011) Climate change and weed adaptation: can evolution of invasive plants lead to greater range expansion than forecasted? Weed Res 51(3):227–240

Cordeau S, Triolet M, Wayman S, Steinberg C, Guillemin JP (2016) Bioherbicides: dead in the water? A review of the existing products for integrated weed management. Crop Prot 87:44–49

Dang YP, Moody PW, Bell MJ, Seymour NP, Dalal RC, Freebairn DM, Walker SR (2015) Strategic tillage in no-till farming systems in Australia's northern grains-growing regions: II. Implications for agronomy, soil and environment. Soil Tillage Res 152:115–123

Dass A, Shekhawat K, Choudhary AK, Sepat S, Rathore SS, Mahajan G, Chauhan BS (2017) Weed management in rice using crop competition-a review. Crop Prot 95:45–52

Davis MB, Shaw RG (2001) Range shifts and adaptive responses to quaternary climate change. Science 292(5517):673–679

De Vita P, Colecchia SA, Pecorella I, Saia S (2017) Reduced inter-row distance improves yield and competition against weeds in a semi-dwarf durum wheat variety. Eur J Agron 85:69–77

Dennis M, Hembree KJ, Bushoven JT, Shrestha A (2016) Growth stage, temperature, and time of year affects the control of glyphosate-resistant and glyphosate-paraquat resistant *Conyza bonariensis* with saflufenacil. Crop Prot 81:129–137

Fausey JC, Renner KA (2001) Environmental effects on CGA-248757 and flumiclorac efficacy/soybean tolerance. Weed Sci 49(5):668–674

Fennimore SA, Cutulle M (2019) Robotic weeders can improve weed control options for specialty crops. Pest Manag Sci

Fennimore SA, Slaughter DC, Siemens MC, Leon RG, Saber MN (2016) Technology for automation of weed control in specialty crops. Weed Technol 30(4):823–837

Fernández-Quintanilla C, Dorado J, Andújar D, Peña JM, Ribeiro Á, de Castro AI, López-Granados F (2018) Recent advances in technologies for a sustainable weed management. In: XVI Congreso de la Sociedad Española de Malherbología, SEMh 2017, Pamplona-Iruña, España, 25-27 octubre, 2017. Universidad Pública de Navarra, pp 375–380

Fernando N, Manalil S, Florentine SK, Chauhan BS, Seneweera S (2016) Glyphosate resistance of C3 and C4 weeds under rising atmospheric CO2. Front Plant Sci 7:910

Fried G, Petit S, Reboud X (2010) A specialist-generalist classification of the arable flora and its response to changes in agricultural practices. BMC Ecol 10(1):20

Ganie ZA, Jugulam M, Jhala AJ (2017) Temperature influences efficacy, absorption, and translocation of 2, 4-D or glyphosate in glyphosate-resistant and glyphosate-susceptible common ragweed (*Ambrosia artemisiifolia*) and giant ragweed (*Ambrosia trifida*). Weed Sci 65(5):588–602

Gharde Y, Singh PK, Dubey RP, Gupta PK (2018) Assessment of yield and economic losses in agriculture due to weeds in India. Crop Prot 107:12–18

Godar AS, Varanasi VK, Nakka S, Prasad PV, Thompson CR, Mithila J (2015) Physiological and molecular mechanisms of differential sensitivity of Palmer amaranth (*Amaranthus palmeri*) to mesotrione at varying growth temperatures. PLoS One 10(5):e0126731

Gourdji SM, Sibley AM, Lobell DB (2013) Global crop exposure to critical high temperatures in the reproductive period: historical trends and future projections. Environ Res Lett 8(2):024041

Harker KN, O'donovan JT (2013) Recent weed control, weed management, and integrated weed management. Weed Technol 27(1):1–11

Harker KN, O'Donovan JT, Smith EG, Johnson EN, Peng G, Willenborg CJ, Gulden RH, Mohr R, Gill KS, Grenkow LA (2015) Seed size and seeding rate effects on canola emergence, development, yield and seed weight. Can J Plant Sci 95(1):1–8

Hatfield JL, Prueger JH (2015) Temperature extremes: effect on plant growth and development. Weather Clim Extremes 10:4–10

Hobbs PR, Sayre K, Gupta R (2007) The role of conservation agriculture in sustainable agriculture. Philos Trans R Soc B Biol Sci 363(1491):543–555

Hulme PE (2017) Climate change and biological invasions: evidence, expectations, and response options. Biol Rev 92(3):1297–1313

Jabran K, Doğan MN (2018) High carbon dioxide concentration and elevated temperature impact the growth of weeds but do not change the efficacy of glyphosate. Pest Manag Sci 74(3):766–771

Jabran K, Mahajan G, Sardana V, Chauhan BS (2015) Allelopathy for weed control in agricultural systems. Crop Prot 72:57–65

Jardine SL, Sanchirico JN (2018) Estimating the cost of invasive species control. J Environ Econ Manag 87:242–257

Johnson GA, Hoverstad TR (2002) Effect of row spacing and herbicide application timing on weed control and grain yield in corn (*Zea mays*). Weed Technol 16(3):548–553

Johnson BC, Young BG (2002) Influence of temperature and relative humidity on the foliar activity of mesotrione. Weed Sci 50(2):157–161

Jursik M, Soukup J, Holec J, Andr J, Hamouzová K (2015) Efficacy and selectivity of pre-emergent sunflower herbicides under different soil moisture conditions. Plant Prot Sci 51(4):214–222

Kaur S, Kaur R, Chauhan BS (2018) Understanding crop-weed-fertilizer-water interactions and their implications for weed management in agricultural systems. Crop Prot 103:65–72

Knezevic SZ, Evans SP, Mainz M (2003) Row spacing influences the critical timing for weed removal in soybean (*Glycine max*). Weed Technol 17(4):666–673

Koger CH, Dodds DM, Reynolds DB (2007) Effect of adjuvants and urea ammonium nitrate on bispyribac efficacy, absorption, and translocation in barnyardgrass (*Echinochloa crus-galli*). I. Efficacy, rainfastness, and soil moisture. Weed Sci 55(5):399–405

Kolb LN, Gallandt ER, Molloy T (2010) Improving weed management in organic spring barley: physical weed control vs. interspecific competition. Weed Res 50(6):597–605

Kolb LN, Gallandt ER, Mallory EB (2012) Impact of spring wheat planting density, row spacing, and mechanical weed control on yield, grain protein, and economic return in Maine. Weed Sci 60(2):244–253

Korres NE, Norsworthy JK, Tehranchian P, Gitsopoulos TK, Loka DA, Oosterhuis DM, Gealy DR, Moss SR, Burgos NR, Miller MR, Palhano M (2016) Cultivars to face climate change effects on crops and weeds: a review. Agron Sustain Dev 36(1):12

Kumaratilake AR, Preston C (2005) Low temperature reduces glufosinate activity and translocation in wild radish (*Raphanus raphanistrum*). Weed Sci 53(1):10–16

Lavorel S, Garnier É (2002) Predicting changes in community composition and ecosystem functioning from plant traits: revisiting the holy grail. Funct Ecol 16(5):545–556

Leguizamón E, Yanniccari M, Guiamet J, Acciaresi H (2011) Growth, gas exchange and competitive ability of *Sorghum halepense* populations under different soil water availability. Can J Plant Sci 91(6):1011–1025

Lemerle D, Lockley P, Koetz E, Diffey S (2013) Herbicide efficacy for control of annual ryegrass (*Lolium rigidum* Gaud.) is influenced more by wheat seeding rate than row spacing. Crop Pasture Sci 64(7):708–715

Lesk C, Rowhani P, Ramankutty N (2016) Influence of extreme weather disasters on global crop production. Nature 529(7584):84

Li Y, Ye W, Wang M, Yan X (2009) Climate change and drought: a risk assessment of crop-yield impacts. Clim Res 39(1):31–46

Liebert JA, Ryan MR (2017) High planting rates improve weed suppression, yield, and profitability in organically-managed, no-till–planted soybean. Weed Technol 31(4):536–549

Liu H, Crawford M, Carvalhais LC, Dang YP, Dennis PG, Schenk PM (2016) Strategic tillage on a Grey Vertosol after fifteen years of no-till management had no short-term impact on soil properties and agronomic productivity. Geoderma 267:146–155

Llewellyn R, Ronning D, Clarke M, Mayfield A, Walker S, Ouzman J (2016) Impact of weeds in Australian grain production. Grains Research and Development Corporation, Canberra, ACT

Lobell DB, Gourdji SM (2012) The influence of climate change on global crop productivity. Plant Physiol 160(4):1686–1697

Lowenberg-DeBoer J, Huang IY, Grigoriadis V, Blackmore S (2018) Economics of robots and automation in field crop production. Precision Agric:1–22

Lubbers MD, Stahlman PW, Al-Khatib K (2007) Fluroxypyr efficacy is affected by relative humidity and soil moisture. Weed Sci 55(3):260–263

Mahan JR, Dotray PA, Light GG (2004) Thermal dependence of enzyme function and inhibition; implications for, herbicide efficacy and tolerance. Physiol Plant 120(2):187–195

Manea A, Leishman MR, Downey PO (2011) Exotic C 4 grasses have increased tolerance to glyphosate under elevated carbon dioxide. Weed Sci 59(1):28–36

Marble SC, Prior SA, Runion GB, Torbert HA (2015) Control of yellow and purple nutsedge in elevated CO2 environments with glyphosate and halosulfuron. Front Plant Sci 6:1

Matzrafi M, Seiwert B, Reemtsma T, Rubin B, Peleg Z (2016) Climate change increases the risk of herbicide-resistant weeds due to enhanced detoxification. Planta 244(6):1217–1227

McFadyen RC, Skarratt B (1996) Potential distribution of *Chromolaena odorata* (Siam weed) in Australia, Africa and Oceania. Agric Ecosyst Environ 59(1–2):89–96

Melland AR, Antille DL, Dang YP (2017) Effects of strategic tillage on short-term erosion, nutrient loss in runoff and greenhouse gas emissions. Soil Res 55(3):201–214

Montzka SA, Dlugokencky EJ, Butler JH (2011) Non-CO_2 greenhouse gases and climate change. Nature 476(7358):43

Moss SR (2010) Non-chemical methods of weed control: benefits and limitations. In: Seventeenth Australasian weeds conference, September, pp 14–19

Mozdzer TJ, Megonigal JP (2012) Jack-and-master trait responses to elevated CO_2 and N: a comparison of native and introduced *Phragmites australis*. PLoS One 7(10):e42794

Neve P, Vila-Aiub M, Roux F (2009) Evolutionary-thinking in agricultural weed management. New Phytol 184(4):783–793

Nguyen TH, Malone JM, Boutsalis P, Shirley N, Preston C (2016) Temperature influences the level of glyphosate resistance in barnyardgrass (*Echinochloa colona*). Pest Manag Sci 72(5):1031–1039

Nguyen T, Bajwa AA, Navie S, O'donnell C, Adkins S (2017) Parthenium weed (*Parthenium hysterophorus* L.) and climate change: the effect of CO_2 concentration, temperature, and water deficit on growth and reproduction of two biotypes. Environ Sci Pollut Res 24(11):10727–10739

Nichols V, Verhulst N, Cox R, Govaerts B (2015) Weed dynamics and conservation agriculture principles: a review. Field Crop Res 183:56–68

Norsworthy JK, Oliveira MJ (2004) Comparison of the critical period for weed control in wide-and narrow-row corn. Weed Sci 52(5):802–807

Oerke EC (2006) Crop losses to pests. J Agric Sci 144(1):31–43

Olson BL, Al-Khatib K, Stahlman P, Isakson PJ (2000) Efficacy and metabolism of MON 37500 in *Triticum aestivum* and weedy grass species as affected by temperature and soil moisture. Weed Sci 48(5):541–548

Ou J, Stahlman PW, Jugulam M (2018) Reduced absorption of glyphosate and decreased translocation of dicamba contribute to poor control of kochia (*Kochia scoparia*) at high temperature. Pest Manag Sci 74(5):1134–1142

Owen MJ, Walsh MJ, Llewellyn RS, Powles SB (2007) Widespread occurrence of multiple herbicide resistance in Western Australian annual ryegrass (*Lolium rigidum*) populations. Aust J Agric Res 58(7):711–718

Patterson DT (1986) Responses of soybean (Glycine max) and three C4 grass weeds to CO_2 enrichment during drought. Weed Sci 34(2):203–210

Patterson DT (1995) Weeds in a changing climate. Weed Sci 43(4):685–700

Pautasso M, Dehnen-Schmutz K, Holdenrieder O, Pietravalle S, Salama N, Jeger MJ, Lange E, Hehl-Lange S (2010) Plant health and global change–some implications for landscape management. Biol Rev 85(4):729–755

Peters K, Breitsameter L, Gerowitt B (2014) Impact of climate change on weeds in agriculture: a review. Agron Sustain Dev 34(4):707–721

Pimentel D, McNair S, Janecka J, Wightman J, Simmonds C, O'connell C, Wong E, Russel L, Zern J, Aquino T, Tsomondo T (2001) Economic and environmental threats of alien plant, animal, and microbe invasions. Agric Ecosyst Environ 84(1):1–20

Pimentel D, Zuniga R, Morrison D (2005) Update on the environmental and economic costs associated with alien-invasive species in the United States. Ecol Econ 52(3):273–288

Pinke G, Blazsek K, Magyar L, Nagy K, Karácsony P, Czúcz B, Botta-Dukát Z (2016) Weed species composition of conventional soyabean crops in Hungary is determined by environmental, cultural, weed management and site variables. Weed Res 56(6):470–481

Refatti JP, Avila LA, Camargo ER, Ziska LH, Oliveira C, Salas-Perez R, Burgos NR (2019) High [CO_2] and temperature increase resistance to cyhalofop-butyl in multiple-resistant *Echinochloa colona*. Front Plant Sci 10:529

Richards CL, Bossdorf O, Muth NZ, Gurevitch J, Pigliucci M (2006) Jack of all trades, master of some? On the role of phenotypic plasticity in plant invasions. Ecol Lett 9(8):981–993

Rosenzweig C, Tubiello FN, Goldberg R, Mills E, Bloomfield J (2002) Increased crop damage in the US from excess precipitation under climate change. Glob Environ Chang 12(3):197–202

Sage RF, McKown AD (2005) Is C4 photosynthesis less phenotypically plastic than C3 photosynthesis? J Exp Bot 57(2):303–317

Sardana V, Mahajan G, Jabran K, Chauhan BS (2017) Role of competition in managing weeds: an introduction to the special issue. Crop Prot 95:1–7

Scheiner SM (2016) Habitat choice and temporal variation alter the balance between adaptation by genetic differentiation, a jack-of-all-trades strategy, and phenotypic plasticity. Am Nat 187(5):633–646

Schwartz LM, Norsworthy JK, Young BG, Bradley KW, Kruger GR, Davis VM, Steckel LE, Walsh MJ (2016) Tall waterhemp (*Amaranthus tuberculatus*) and palmer amaranth (*Amaranthus palmeri*) seed production and retention at soybean maturity. Weed Technol 30(1):284–290

Schwartz-Lazaro LM, Green JK, Norsworthy JK (2017) Seed retention of palmer amaranth (*Amaranthus palmeri*) and Barnyardgrass (*Echinochloa crus-galli*) in soybean. Weed Technol 31(4):617–622

Shaner DL, Beckie HJ (2014) The future for weed control and technology. Pest Manag Sci 70(9):1329–1339

Sharma SD, Singh M (2001) Environmental factors affecting absorption and bio-efficacy of glyphosate in Florida beggarweed (*Desmodium tortuosum*). Crop Prot 20(6):511–516

Smith RG, Mortensen DA, Ryan MR (2010) A new hypothesis for the functional role of diversity in mediating resource pools and weed–crop competition in agroecosystems. Weed Res 50(1):37–48

Soltani N, Dille JA, Burke IC, Everman WJ, VanGessel MJ, Davis VM, Sikkema PH (2016) Potential corn yield losses from weeds in North America. Weed Technol 30(4):979–984

Soltani N, Dille JA, Burke IC, Everman WJ, VanGessel MJ, Davis VM, Sikkema PH (2017) Perspectives on potential soybean yield losses from weeds in North America. Weed Technol 31(1):148–154

Soltani N, Dille JA, Gulden RH, Sprague CL, Zollinger RK, Morishita DW, Lawrence NC, Sbatella GM, Kniss AR, Jha P, Sikkema PH (2018) Potential yield loss in dry bean crops due to weeds in the United States and Canada. Weed Technol 32(3):342–346

Taub DR, Miller B, Allen H (2008) Effects of elevated CO_2 on the protein concentration of food crops: a meta-analysis. Glob Chang Biol 14(3):565–575

Trenberth KE (2011) Changes in precipitation with climate change. Clim Res 47(1–2):123–138

Turner KG, Fréville H, Rieseberg LH (2015) Adaptive plasticity and niche expansion in an invasive thistle. Ecol Evol 5(15):3183–3197

Utstumo T, Urdal F, Brevik A, Dørum J, Netland J, Overskeid Ø, Berge TW, Gravdahl JT (2018) Robotic in-row weed control in vegetables. Comput Electron Agric 154:36–45

Valerio M, Tomecek MB, Lovelli S, Ziska LH (2011) Quantifying the effect of drought on carbon dioxide-induced changes in competition between a C3 crop (tomato) and a C4 weed (*Amaranthus retroflexus*). Weed Res 51(6):591–600

Varanasi A, Prasad PV, Jugulam M (2016) Impact of climate change factors on weeds and herbicide efficacy. Adv Agron 135:107–146

Vougioukas SG (2019) Agricultural robotics. Annu Rev Control Robot Auton Syst

Walsh MJ, Powles SB (2014) High seed retention at maturity of annual weeds infesting crop fields highlights the potential for harvest weed seed control. Weed Technol 28(3):486–493

Walsh M, Ouzman J, Newman P, Powles S, Llewellyn R (2017) High levels of adoption indicate that harvest weed seed control is now an established weed control practice in Australian cropping. Weed Technol 31(3):341–347

Waltz AL, Martin AR, Roeth FW, Lindquist JL (2004) Glyphosate efficacy on velvetleaf varies with application time of day. Weed Technol 18(4):931–939

Waryszak P, Lenz TI, Leishman MR, Downey PO (2018) Herbicide effectiveness in controlling invasive plants under elevated CO2: sufficient evidence to rethink weeds management. J Environ Manag 226:400–407

Weedsmart (2017) SwarmFarm: targeting small weeds all year. https://weedsmart.org.au/swarmfarm-targeting-small-weeds-all-year/

Westwood JH, Charudattan R, Duke SO, Fennimore SA, Marrone P, Slaughter DC, Swanton C, Zollinger R (2018) Weed management in 2050: perspectives on the future of weed science. Weed Sci 66(3):275–285

Wiggins MS, McClure MA, Hayes RM, Steckel LE (2015) Integrating cover crops and POST herbicides for glyphosate-resistant palmer amaranth (*Amaranthus palmeri*) control in corn. Weed Technol 29(3):412–418

Wijffels S, Roemmich D, Monselesan D, Church J, Gilson J (2016) Ocean temperatures chronicle the ongoing warming of earth. Nat Clim Chang 6(2):116

Young SL, Pitla SK, Van Evert FK, Schueller JK, Pierce FJ (2017) Moving integrated weed management from low level to a truly integrated and highly specific weed management system using advanced technologies. Weed Res 57(1):1–5

Yu Q, Cairns A, Powles S (2007) Glyphosate, paraquat and ACCase multiple herbicide resistance evolved in a *Lolium rigidum* biotype. Planta 225(2):499–513

Zhang W, Webster EP, Selim HM (2001) Effect of soil moisture on efficacy of imazethapyr in greenhouse. Weed Technol 15(2):355–359

Zhu C, Zeng Q, Ziska LH, Zhu J, Xie Z, Liu G (2008) Effect of nitrogen supply on carbon dioxide–induced changes in competition between rice and barnyardgrass (*Echinochloa crus-galli*). Weed Sci 56(1):66–71

Ziska LH (2010) Elevated carbon dioxide alters chemical management of Canada thistle in no-till soybean. Field Crop Res 119(2–3):299–303

Ziska LH (2016) The role of climate change and increasing atmospheric carbon dioxide on weed management: herbicide efficacy. Agric Ecosyst Environ 231:304–309

Ziska LH, Teasdale JR (2000) Sustained growth and increased tolerance to glyphosate observed in a C3 perennial weed, quackgrass (*Elytrigia repens*), grown at elevated carbon dioxide. Funct Plant Biol 27(2):159–166

Ziska LH, Faulkner S, Lydon J (2004) Changes in biomass and root: shoot ratio of field-grown Canada thistle (*Cirsium arvense*), a noxious, invasive weed, with elevated CO_2: implications for control with glyphosate. Weed Sci 52(4):584–588

Climate Change: Impact on Plant Pathogens, Diseases, and Their Management

Mandeep Singh Hunjan and Jagjeet Singh Lore

Introduction

Anthropogenic emissions (combustion of fossil fuels resulting in the release of carbon dioxide [CO_2] as well as the release of other gases such as methane) and anthropological shifts in demographics, economics, technology and social behaviour have contributed significantly to climate change. With earth's temperatures that have already been increased by more than 1 °C since the 1880s, the Intergovernmental Panel on Climate Change (IPCC) has projected that global temperatures will have a further increase of 1.5 °C between 2030 and 2052 (IPCC 2014). The IPCC further predicts that increase in global mean temperature from 1 to 3 °C above 1990 levels will have either positive or negative impacts depending upon the region, with net annual cost set to surge as the temperature increases. The shared interactions between plant hosts, pathogens and their environment in causing plant diseases are described by the disease triangle (Garrett et al. 2006; Grulke 2011; Nazir et al. 2018), with environment playing a key role to determine the outcome of these interactions. Changes in climate such as increase in temperature and atmospheric CO_2 and the frequency and intensity of extreme weather fluctuations such as drought and flooding affect host plant resistance to pathogens (Dossa et al. 2015). The changes in these climate variables may reshape host-pathogen interactions and influence spatial and temporal development of disease epidemics (Chakraborty 2005; Burdon et al. 2006; Garrett et al. 2006; Crowl et al. 2008; Eastburn et al. 2011). Understanding the effects of climate change on disease dynamics is crucial in adopting appropriate control measures and identifying sources of resistance to diseases (Chakraborty and Pangga 2004; Ghini et al. 2008). Weather is a key driver of endemic bacterial dis-

M. S. Hunjan · J. S. Lore (✉)
Punjab Agricultural University, Ludhiana, India
e-mail: jagjeetsingh-pbg@pau.edu

K. Jabran et al. (eds.), *Crop Protection Under Changing Climate*,
https://doi.org/10.1007/978-3-030-46111-9_4

eases and the transient temperature shifts predicted to be increased during this century; these have potential to either suppress or cause disease outbreaks (Anderson et al. 2004; Fischer and Knutti 2015).

Although the outcome of host-pathogen interactions is driven by a number of different climate variables, temperature is the most important factor. Increase in temperature may result in greater crop vulnerability to disease (Coakley et al. 1999) due to the interactive effect of temperature on both the host plant and pathogen. In addition, elevated CO_2 has also shown substantial effect on disease development under controlled conditions. Most diseases studied by Chakraborty and Pangga (2004) increased in severity under the CO_2-enriched environments that resulted in alterations to the host-pathogen relationship. Within this context, we aim to consolidate information on the effects of different drivers of climate change on different groups of pathogens, on host-pathogen interactions and on disease management strategies with a primary focus on two key drivers: elevated CO_2 and temperature.

Impact of Climate Change on Plant Pathogens

A disease occurs only when a virulent pathogen attacks a susceptible host under suitable environmental conditions at an appropriate time (Agrios 2004). Climate change can change host resistance towards pathogens by altering host physiology and pathogen aggressiveness. Increases in global temperature will shift agroclimatic zones towards the poles, thereby changing the geographic distribution of current diseases by introducing pathogens into new areas. The quick acclimatization of pathogens in a region depends upon how the pathogen disperses, how it survives in the off-season and its ability to adapt to any change in its host's biology in a new environment. Concurrently, new pathogens and diseases may emerge in existing agroecosystems. More aggressive, climate-resilient strains of existing pathogens may emerge. The impact on different groups of pathogens may vary depending upon their biology, level of host-specialization, survival and dispersal modes and ability to cope with climate change. Here we discuss the impact of climate change on major groups of plant pathogens, viz. fungal, bacterial and viral.

Impact on Fungal Pathogens Fungi are the most dominant group of plant pathogens, displaying diverse modes of parasitic interaction with their host plants. Fungal pathogenesis is greatly influenced by prevailing atmospheric conditions, particularly temperature and moisture. Temperature is a critical factor during different life stages of a fungal pathogen, and any change in temperature will significantly influence fungal reproduction, rate of infection, number of infection cycles, long and short distance dispersal as well as off-season survival. Change in temperature might lead to activation of dormant races of pathogens, thus leading to sudden outbreak of an epidemic. Increase in temperature coupled with high soil moisture creates a warm and humid climate highly favourable to soilborne pathogens and foliar diseases. For example, increase in ambient temperature has increased potential for rice blast (*Magnaporthe grisea*) incidence and epidemics for both cooler and subtropical rice-growing regions. Similarly, the severity of *Septoria* leaf spot (*Septoria lycopersici*) and stem rust (*Puccinia graminis*) in oat cultivars has also increased due to temperature sensitivity. Using a free-air CO_2 enrichment, Kobayashi et al. (2006) showed that elevating CO_2 by ~200–280 μmol mol^{-1} above ambient levels poses a potential risk for higher infection of leaf blast and sheath blight of rice in Japan.

Temperature regulates the reproduction rate of many plant pathogenic fungi (Legler et al. 2012). Elevated temperature favours accelerated spore germination in *Puccinia substriata* as well as the reproduction rate of *Monosporascus cannonballus*. Temperature increase will introduce the quick evolution of fungal pathogens due to longer seasons. It will also lead to the development of more aggressive fungal strains due to hastened gene recombination through overwintering sexual stages.

In recent years, changes in disease scenario have affected wheat production and have caused severe yield losses and reduced grain quality. Outbreak of yellow rust in the northern Indian state of Punjab in recent years is attributed to the new pathotype 78S84 which can cause infection at higher temperatures. In recent years, wheat yellow rust has started appearing in late December due to favourable weather conditions of increased temperatures (Prashar et al. 2007; Jindal et al. 2012). The incidence of powdery mildew and foliar blights has also increased in northern regions of India. The UG99 race of stem rust, a new threat to wheat production, has entered the Middle East and presents the potential for wide-scale yield losses. Powdery mildew has been predicted to cause serious losses in winter wheat in China under simulated elevated temperature conditions (Tang et al. 2017). Additionally, rice blast disease was not present in Punjab before the year 2000; however, neck blast has become a serious problem in Basmati rice in recent years.

Elevated CO_2 levels in the atmosphere cause physiological changes in plant morphology such as increases in leaf size, leaf thickness, and number of leaves (Pritchard et al. 1999). A thick canopy leads to increased duration of leaf surface moisture, presenting a highly favourable environment for spore germination and penetration in foliar pathogens leading to disease development (Garrett et al. 2006). Deutsch et al. (2008) suggested that warming may enhance fitness of pathogens in their environment on the host causing increased risk of disease epidemic outbreak. Elevated CO_2 levels (780 ppm) have been shown to increase the susceptibility of wheat variety *Remus* to *Fusarium* head blight and *Septoria tritici* blotch (Vary et al. 2015). In contrast, elevated concentrations of CO_2 and O_3 reduced downy mildew (*Peronospora manshurica*) severity but increased brown spot (*Septoria glycines*) severity in soybean (Eastburn et al. 2010). Elevated CO_2 and O_3 levels have also been shown to increase resistance in barley against powdery mildew (*Blumeria graminis*) through formation of papillae and silicon accumulation at sites of appressorial penetration. Greater fecundity in some pathogens as a result of increased levels of atmospheric CO_2 could lead to polycyclic epidemics in crops (Chakraborty and Datta 2003). It is difficult to generalize the effect of elevated CO_2 on a particular pathogen as it also depends on the effect of elevated CO_2 on the host plant, suggesting that the effect can be either positive or negative (Gautam et al. 2013).

Climate change leads to an increase in tropospheric ozone that hinders plant photosynthesis, resulting in poor growth due to water-soaked lesions. This also predisposes plants to biotic attack, particularly in the presence of necrotrophic and root-infecting fungi. It has been shown that *Botrytis cinerea* causing grey mould is more aggressive on onions exposed to ozone (Wukasch and Hofstra 1977).

Impact on Bacterial Pathogens For every disease to occur, there is optimal temperature range, e.g. *Xanthomonas oryzae* pv. *oryzae* (*Xoo*), that fails to efficiently colonize rice xylem when the daytime temperature exceeds 35 °C (Horino et al. 1982). *Ralstonia solanacearum*, a major cause of bacterial wilt and brown rot in potato, fails to survive the high summer temperatures of Punjab, India. More than 40 genera of bacteria are reported to be associated with plants as pathogens. The emergence of a number of bacterial genera as a serious problem worldwide could possibly be due to global warming. Examples include *Acidovorax avenae* subsp. *avenae* infecting upland rice in southern Europe, *Burkholderia andropogonis* on jojoba in eastern Australia, *B. glumae* on rice in the southern United States and *Dickeya zeae* in rice in north India especially under high temperatures. Elevated temperature interferes with bacterial physiology, genetics and bacterial-plant interactions.

Exposure to high temperatures may increase the fitness of plant pathogenic bacteria on their respective host. Plant pathogenic bacteria, especially the xanthomonads and pseudomonads, produce copious amounts of extracellular polysaccharides (EPS) during their interaction with the host plant. This production is usually triggered by biotic and abiotic signals. Under natural conditions, bacteria use EPS to aggregate themselves and to make biofilms in order to protect themselves from unfavourable environmental conditions, antimicrobial compounds, etc. In addition, bacteria also use EPS for communication, adhesion, energy storage, etc. (Wingender et al. 1999; Vardharajula and Ali 2015). Exposure to high temperatures has been shown to trigger evolution of mutant strains with elevated levels of EPS (Nandal et al. 2005). High EPS producer strains of *Xoo* have been shown to be highly aggressive in rice (Kumar and Sakthivel 2001; Hunjan et al. 2014). High temperature (34.5 °C) attenuated the virulence of soft rot bacterium (*Erwinia carotovora* subsp. *carotovora*), but not in the particular strain EC153 that produced high levels of rRNA, N-acyl homoserine lactone and extracellular proteins causing extensive maceration of celery petioles and Chinese cabbage (Hasegawa et al. 2005).

Temperature also affects expression of effector genes in plant pathogenic bacteria and their recognition events in host plants during pathogenesis. Pathogen-associated molecular patterns (PAMPs) are conserved microbial signatures that evoke basal or PAMP-triggered immunity (PTI) in hosts upon recognition by PAMP recognition receptors in the host. PTI has been shown to be increased when *Arabidopsis* was exposed to 28 °C for short periods (Cheng et al. 2013), whereas *R* gene-based effector-triggered immunity (ETI) through *Arabidopsis RPM1*, *RPS2* and *RPS4* against *Pseudomonas syringae* is compromised at temperatures above 30 °C (Wang et al. 2009). An exception to this is rice *R* gene *Xa7*, which is more effective at high temperature against bacterial blight disease (Webb et al. 2010).

Virulence of *Agrobacterium* strains on *Kalanchoe* is attenuated at or above 32 °C possibly due to reduced *Vir* gene expression (Jin et al. 1993). The level of accumulation of virulence proteins in *Agrobacterium tumefaciens* was strongly reduced, and T-pilus assembly failed to occur at 28 °C as compared to 26 °C (Baron et al. 2001).

Impact on Viral Pathogens and Their Vectors Viral pathogens infect their plant hosts through their close association with vectors, thereby adding a new dimension to the disease triangle. The outbreak of a virus disease epidemic is therefore limited by the climatic requirements of virus vectors (Malmstrom et al. 2011). Climate change may affect both host plants and the type of vectors associated with them, determining the severity of its impact on viruses associated with that host (Jones 2009). It may also affect the range of vectors, their migration and biology (Canto et al. 2009). Jones and Barbetti (2012) have authored a comprehensive review where they tried to establish the possible effects of direct and indirect climate change parameters on the many vectors, viruses and host factors. Migrations of viruses such as potato leaf roll virus (PLRV) and potato yellow vein virus (PYVV) that are adapted to warmer regions are common at higher altitudes (Jones 2014). Elevated temperatures hastened yellow dwarf disease development and symptom expression in wheat infected with barley yellow dwarf virus-PAV (BYDV-PAV). BYDV-PAV inoculated wheat seedlings grown at higher temperatures (10.0–21.1 °C, night-day) showed higher titre than those grown at ambient (5.0–16.1 °C, night-day) temperatures (Nancarrow et al. 2014). In another study, elevated CO_2 (650 μmol mol^{-1}) was also shown to aggravate yellow dwarf disease and BYDV-PAV titre in wheat (Trebicki et al. 2015). High virus titres may not always result in enhanced symptom expression. Del Torro et al. (2015) reported that *Nicotiana benthamiana* plants inoculated with cucumber mosaic virus, PVY or potato virus X under elevated CO_2 did not affect symptom expression although virus titre was high. Elevated CO_2 reduced disease development and enhanced plant resistance in tobacco plants against potato virus Y (Matros et al. 2006) and in tomato against tobacco mosaic virus (Zhang et al. 2015) and tomato yellow leaf curl virus (Huang et al. 2012). Elevated temperature and CO_2 can significantly alter plant biochemistry, and hence plants may respond differentially to express defence responses towards insect vectors of viruses. It can affect insect fecundity, feeding rates, off-season survival and dispersal (Trebicki et al. 2017).

Most of the published reports on aphids have suggested that aphids will react strongly to environmental changes. Aphids have shorter generation times, and an increase in the number of virus carrying populations will eventually lead to higher viral load *in planta*. The aphid transmissible complex of BYDV in wheat and PVY in potato are amenable to show significant effects on the prevalence of infection because of elevated temperature and CO_2. In mild winters, high levels of aphid movement during spring have resulted in greater levels of PVY infection in potatoes. Greenhouse adapted viruses such as the pepino mosaic virus (PepMV) have moved out and became predominant in field conditions due to climate change. Natural climatic barriers in more temperate regions are deteriorating under changes in early winter temperatures, allowing for the natural spread of diseases, pests and vectors at prolific rates. Increased likelihood for the outbreak of plant disease vectors such as aphids, whiteflies, thrips and beetles may cause severe epidemics in these regions.

Impact of Climate Change on *R* Gene-Mediated Disease Control The defence mechanism of *R* genes is considered the major source of resistance in various crop-breeding programs. Pathogen populations are subjected to selection pressure due to changing environmental conditions, thus affecting the durability of a deployed *R* gene. *R* genes can be sensitive to temperature and are only effective at either high or low temperatures (Dyck and Johnson 1983; Eizenberg et al. 2003; Webb et al. 2010; Dossa et al. 2015).

The wheat stripe rust resistance gene *Yr36* imparts resistance to a wide spectrum of *Puccinia striiformis* f. sp. *tritici (Pst)* races. The gene is reported to be effective at high temperatures (25–35 °C) although unable to resist fungal attack at temperatures below 15 °C (Uauy et al. 2005). Understanding the interconnectivity of environmental factors, pathogen evolution and *R* gene effectiveness requires knowledge of *R* gene function mechanisms and how a particular function is affected by various weather parameters. Loss of function in an effector protein through a single mutation may lead to loss of recognition by plant R proteins, resulting in evolution of new races or widening of the pathogen host range. In another context, a mutation in the effector protein can also result in reduced pathogenic fitness (Burdon et al. 2006). Effectiveness of bacterial blight resistance gene *Xa7* has been correlated with mutations in *Xoo* that resulted in reduced pathogenic fitness of the bacterium due to its reduced aggressiveness on susceptible host cultivars. These mutations in *Xoo* have occurred specifically in the pathogen effector gene *avra7*, and thus selection pressure imposed on the bacterium population by the *Xa7* gene resulted in reduced pathogenic fitness (Vera Cruz et al. 2000; Webb et al. 2010). The pathogen effectors block MAPK-mediated signalling pathways invoked by PRRs, however incite *R* gene-mediated effector-triggered immunity (ETI).

Elevated temperatures affect ETI in many pathosystems. At lower temperatures (10–23 °C), ETI signalling is preferentially activated, whereas PTI signalling is activated at higher temperatures (23–32 °C). The tobacco *N* gene against tobacco mosaic virus (Samuel 1931); tomato *Cf* against *Cladosporium fulvum* (de Jong et al. 2002); *Arabidopsis RPW8* against powdery mildew (Xiao et al. 2003); *Arabidopsis RPM1*, *RPS2*, and *RPS4* genes against *P. syringae*; or potato *Rx* fail to impart ETI at elevated temperatures (Wang et al. 2009). Although elevated temperature could inhibit ETI, *Xa7*, a rice disease resistance protein against *Xoo*, is more effective at higher temperatures (Webb et al. 2010).

Temperature also affects the pathogen fitness traits as well as genotype-environment interactions in case of *Pst* causing stripe or yellow rust of wheat. The pathogen shows increased aggressiveness on exposure to high temperatures, suggesting that rust fungi can adapt to warmer climatic conditions (Mboup et al. 2012). Some of the *R* genes are effective either at high or low temperatures. During a study conducted at Punjab Agricultural University, India), it was found that high temperature decreases the effectiveness of the bacterial blight resistance gene *Xa4*, while

Xa7, another bacterial blight resistance gene, was effective at high temperatures. The phenotypic results were corroborated by reduced production of different defence-related proteins in rice near isogenic line IRBB-4 at high temperatures. Hartleb and Heitefuss (1997) suggested that disease resistance can also be dependent on temperature, for example, phoma stem canker (*Leptosphaeria maculans*) of rapeseed mustard where resistance is expressed at 15 °C but not at 25 °C. Cohen et al. (2017) found certain differentially expressed genes as temperature responsive in rice cultivars containing *Xa7* using RNA-Seq technology. They suggested that plant hormone abscisic acid plays an important role in response to high temperature and pathogen attack. Tolerance to abiotic stresses may also induce enhanced resistance to biotic stresses.

Development of New Races and Diseases in an Ecosystem

Pathogens and their plant hosts are forever engaged in a co-evolutionary arms race. Pathogens evolve to evade recognition by plant hosts, while the latter evolve to recognize a wide array of pathogen effectors. To evolve new pathogenicity traits, pathogens undergo horizontal gene transfer where new mutations arise and the gene pool is widened through crossing of genetically diverse individuals leading to natural selection and subsequent adaptation of new variants. Plant pathogens are spreading globally with latitudinal shifts in their distribution for the last 60 years, largely in line with climate change projections. The damage imposed by these organisms may have serious consequences on economies of the developing world because of greater dependence on agriculture in these countries, as well as limited resources and technological interventions to manage crop diseases (Bebber et al. 2014).

Pst, the wheat pathogenic fungi with global prevalence and huge potential for long-distance migration, poses a serious invasion threat to new areas (Dean et al. 2012; Beddow et al. 2015; Ali et al. 2014; Chen et al. 2014). Often considered as a cold-loving pathogenic fungi (Brown and Hovmoller 2002), it has recently been able to invade wheat grown in warmer regions, demonstrating the thermal adaptation in it. New aggressive races of *Pst* (PstS1/S2) possessing broader virulence spectra suited to higher temperatures have been described (Milus et al. 2009; Walter et al. 2016). Incursion of a close lineage of *PstS2* was also observed in Mediterranean growing regions in 2004 (Bahri et al. 2009). Additionally, a warm temperature-loving race of *Pst* 78S84 broke down the resistance of *R* genes *Yr9* and *Yr27* in wheat mega variety PBW 343 in northwestern India.

Lyon and Broders (2017) stated that temperature and precipitation affect spore germination, infection and survival of *Pst.* Although *Pst* prefers temperatures between 7 and 12 °C, (Chen et al. 2014), temperature range adaptation has been observed among isolates (Sharma-Poudyal et al. 2014). Experiments conducted at increased temperature (18 °C) revealed that races PstS1/S2 had shorter latent periods and higher germination rates as compared to pre-2000 US *Pst* isolates (Milus et al. 2006, 2009; Walter et al. 2016). The post-2000 *Pst* populations were genetically distinct (Ali et al. 2014) and coupled with phenotypic plasticity. This genetic differentiation might have resulted in their high temperature adaptation. Similarly, local thermal adaptation was also reported in French (Mboup et al. 2012) and Australian (Loladze et al. 2014) isolates. The warrior race of *Pst* that invaded northwestern Europe in 2011 showed wider adaptation to a range of temperatures due to its presence in both warm (Spain) and cold (Sweden) areas (Hovmoller et al. 2016). The strains of Warrior were categorized as thermal generalists as they showed the highest infection efficiency under optimal temperatures of 10 °C and 15 °C (de Vallavielle-Pope et al. 2018).

Mariette et al. (2016a) reported local thermal adaptation of *Phytophthora infestans* isolates in different climatic zones of western Europe. Their findings challenge the common notion that invasive behaviour is linked to increased aggressiveness for this pathogen (Mariette et al. 2016b). Their observations also predicted stable or reduced risk of late blight epidemics in future decades as already hypothesized by Sparks et al. (2014).

In the last century, Panama disease (*Fusarium oxysporum* f. sp. *cubense* Race 1) wiped out the banana industry in Central America. However, this was overcome by planting Cavendish bananas. However, Cavendish bananas succumbed to new race FocTR4 of this pathogen in Laos, Myanmar and Vietnam which has since spread from Southeast Asia across the Indian subcontinent, the Middle East and into Africa (Ordonez et al. 2015; Zheng et al. 2018). Climate change provided favourable temperature conditions and increased canopy wetness for spore germination and growth of *Pseudocercospora fijiensis*, the causal agent of black Sigatoka disease. The disease has been described as an emerging threat to banana cultivation (Churchill 2011) and is responsible for huge yield losses in banana plantations worldwide (Bebber 2019). Bebber (2019) parameterized an infection model on this disease that revealed an increased (44.2%) risk of infection in banana-growing areas of Latin America and the Caribbean since the 1960s.

Spatial and temporal dynamics of plant diseases are poorly understood, although models to forecast the impact of climate on crop production have been around for the last 40 years. Bregaglio et al. (2013) applied current climate as baseline and projected their simulations on potential infection events of fungal infections of wheat, rice and grapes in Europe. They forecasted brown rust (*Puccinia recondita*) to increase in wheat by 20–100%, whereas yellow rust was projected to increase by 5–20% in cold areas. Rice blast (*Pyricularia oryzae*) and brown spot (*Bipolaris oryzae*) were postulated to increase in all European rice-growing areas, specifically in northern Italy by almost 100%. Infection of *Plasmopara viticola*, the causal agent of grapes downy mildew, was estimated to increase by 5–20% throughout Europe. Their findings suggested presumed disease pressure on crops under changing climatic conditions and eventually the future challenges that farmers will face.

Impact of Climate Change on Disease Management Practices

Changing Planting Date Climate change may influence the sowing or planting date of many crops. Cultural practices and changing weather patterns influenced *Septoria tritici* blotch (STB) caused by *Zymoseptoria tritici*. The severity of the disease was decreased with low temperatures at −2 °C in the early stages of crop growth. Stewart's wilt, (*E. stewartii*), vectored by the corn flea beetle (*Chaetocnema pulicaria*) is highly dependent upon the survival of the beetle through winter. A forecast model based on winter temperatures predicted that the survival of vector is higher in warmer winters (Petzoldt and Seaman 2006).

Altering Fungicide Schedule *Septoria tritici* blotch (STB) is the major pathogen problem for wheat in the United Kingdom. Climate conditions have changed from the last decade and efforts have been made for forecasting of STB. The decision system for the timing of fungicide application has been made based upon different climate variables. There is a need to evaluate new broad-spectrum multi-target site antifungal treatments. In north-east of the United States, it was predicted that for each 1 °C rise in temperature, potato late blight would occur 4–7 days earlier, and the susceptibility period would be increased by 10–20 days. There may be a need for adding 1–4 fungicide foliar applications which would increase both farmer costs and environmental risks. In northern latitudes, a higher number of fungicide applications are required under frequent rainfall causing difficulty in retention of contact fungicides on the plant (Wolfe et al. 2008). The introduction of new fungicides with a larger effect under rainfall conditions might help to minimize this problem (Hannukkala et al. 2007). In response to higher CO_2 and elevated temperature, plant morphological characteristics like smaller stomatal openings or thicker epicuticular waxes on the leaves could reduce or delay the uptake and translocation of systemic fungicides (Juroszek and Tiedemann 2011). The efficacy of fungicides can be increased with proper timing of fungicide application (Bedos et al. 2002). Stem rot (*Sclerotium rolfsii*) of peanut managed by applying fungicide early in the morning improved spray deposition in the lower canopy of the plant (Augusto et al. 2010). The higher number of fungicide applications could be needed to control problematic diseases under high CO_2 and increased temperature. Additional fungicide sprays are required as rice plants become increasingly susceptible to leaf blast under higher CO_2 concentrations (Kobayashi et al. 2006). It was predicted that two extra applications of fungicide might become essential in order to manage downy mildew of grapevine under elevated temperatures in the northwestern part of Italy by the end of the twenty-first century (Juroszek and Tiedemann 2011).

Efficacy of Fungicides A high range of temperature and CO_2 concentrations support vegetative growth of plants (Gutierrez et al. 2008) that lowers the availability of pesticide per unit area in plants. It is accompanied by reduced pesticide uptake by the roots from the upper soil layer as they penetrate deeper soil layers. Temperature increase, reduced rainfall and high concentrations of atmospheric CO_2 can develop fungicide resistance in pathogens resulting in an increased number of fungicide applications (Delcour et al. 2015).

Fungicide uptake and translocation in plants are influenced by precipitation and will be limited under dry conditions due to a reduced rate of transpiration (Keikotlhaile 2011).

Impacts of Climate Change on Food Safety Threats to food safety posed by the climate change can be predominantly categorized into three:

(i) **Food-borne contamination:** It has been reported that a warmer climate in combination with inappropriate food handling may contribute to increased incidences of food-borne diseases (IPCC 2014). In many countries, the main food-borne pathogens are *Salmonella*, *Campylobacter* and *Escherichia coli.* In temperate countries, strong seasonal patterns to the incidence of food-borne diseases have been observed (Seguin 2008). The links between ambient temperature and food-borne pathogen infections have been demonstrated (Seguin 2008).

(ii) **Mycotoxin contamination of food grains:** Mycotoxins are toxins produced by *Alternaria*, *Aspergillus*, *Fusarium* and *Penicillium.* Probably the most commonly known mycotoxin is the highly carcinogenic aflatoxin, which is produced by *Aspergillus flavus.* Mycotoxins are highly dependent on appropriate temperatures and water availability. Cool and temperate climates result in increased presence of aflatoxin due to increased *Aspergillus* occurrence, while tropical climates may experience a decline due to unsuitable high temperatures for *Aspergillus* (Paterson and Lima 2010). The major problem with the interactive effect of climate change on fungal growth and mycotoxin production is that mycotoxins may contaminate staple cereals such as wheat or corn (Medina 2017) which have an enormous importance for food security.

(iii) **Contaminants and residues:** Climate change will also affect the contamination of food sources with chemicals, such as plant protection product residues. As already discussed in this chapter, the increased use of pesticides on crops may lead to excessive residues and acute food safety risks.

Conclusions and Probable Strategies to Mitigate Effect of Climate Change on Plant Diseases

Over the next 100 years, global temperatures are expected to rise along with increasing levels of atmospheric O_3 and CO_2. Severe weather events like heat waves, erratic rainfall patterns and drought will probably become more common. Due to temperature rise, there may be an increase in crop production because of longer growing seasons in temperate regions. Simultaneously, climate change will affect diseases. There will be an increase in plant disease intensity, frequency of outbreaks and the introduction of pathogens to new areas. According to Helfer (2014), increases in temperature had a beneficial effect on survival of the wheat leaf rust pathogen depending on the availability of leaf wetness and humidity. Contrary to this, the effect of increased CO_2 varied among susceptible cultivars, while increased O_3 had a negative effect. Stem rust resistance due to *R* gene *Sr31* is also in danger under threat of the Ug99 race of stem rust caused by *P. graminis* f. sp. *tritici* due to climate change. There is a threat perception of potato late blight and major diseases of rice, namely, sheath blight (*R. solani*) and blast (*P. oryzae*) due to elevated CO_2 concentrations and high temperatures. In arid, hot conditions, incidence of *M. phaseolina* is higher, and there are chances for the introduction of this pest to new areas under changing climatic conditions (Fones and Gurr 2017).

The geographical and temporal distribution of diseases and management methods will have to be modified according to climate change scenarios. The fungicide residue dynamics in foliage can be altered by temperature change and variable rainfall. The degradation of products can also be altered. Plant physiology or morphology changes because of high concentration of CO_2 in the atmosphere and variations in precipitation and temperature. Therefore, it is important to devise strategies for plant disease management in respect to climate change. The major consequence of climate change in the pathogen-host interaction is genetic resistance to diseases in plant. Plant physiological changes lead to modification in the resistance mechanisms of plant cultivars obtained as a result of genetic engineering and traditional methods. Current disease resistance breeding programs should include traits for tolerance to abiotic factors such as heat stress, elevated CO_2 and water stress. In addition, genome-wide association mapping can be used to understand and target quantitative trait loci for complex traits for abiotic and biotic stresses.

New climate-resilient strains of biocontrol agents with high competitive ability to survive in introduced environment may be needed. Region- and crop-specific strains may have to be incorporated and integrated into pest management approach. IPM is an ecosystem-based strategy, which emphasizes the long-term prevention of pests or their damage through a combination of methods such as cultural practices, habitat management, use of resistant varieties and biological control (Strand 2000). Detailed evaluation of different cropping systems can suggest changes in planting dates and cultural and other agronomic practices to avoid coinciding the susceptible stage of hosts with the virulent phase of the pathogen. While well-developed models for major crops exist, in the case of plant diseases, models are limited to a few major pathogens (Newbery et al. 2016). De Wolf and Isard (2007) suggested disease-forecasting models with improved quality are needed to guide farmers. This type of prediction tool may allow farmers to respond timely and efficiently with public-private partnership. Under changing climatic conditions, crops and varieties may spread to regions and locations where they have not been grown previously.

It is important to understand the potential effects of climate change on disease epidemics on a spatial and temporal scale. However, due to lack of multifactor plant disease simulation models, it is difficult to predict how climate variability will affect disease development and their management. In general, high temperatures and an increased concentration of atmospheric CO_2 will lead to changes in pathogen populations as well as an increased level of susceptibility to plant diseases.

References

Agrios GN (2004) Plant pathology, 5th edn. Elsevier, London, 922p

Ali S, Gladieux P, Leconte M, Gautier A, Justesen AJ, Hovmoller MS, Enjalbert J, de Vallavieille-Pope C (2014) Origin, migration routes and worldwide population genetic structure of the wheat yellow rust pathogen *Puccinia striiformis* f. sp. *tritici*. PLoS Pathog 10:e1003903

Anderson PK, Cunningham AA, Patel NG, Morales FJ, Epstein PR, Daszak P (2004) Emerging infectious diseases of plants: pathogen pollution, climate change and agrotechnology drivers. Trends Ecol Evol 19:535–544

Augusto J, Brenneman TB, Culbreath AK, Sumner P (2010) Night spraying peanut fungicides II. Application timings and spray deposition in the lower canopy. Plant Dis 94:683–689

Bahri B, Leconte M, Ouffroukh A, de Vallavieille-Pope C, Enjalbert J (2009) Geographic limits of a clonal population of wheat yellow rust in the Mediterranean region. Mol Ecol 18:4165–4179

Baron C, Domke N, Beinhofer M, Hapfelmeier S (2001) Elevated temperature differentially affects virulence, VirB protein accumulation, and T-pilus formation in different *Agrobacterium tumefaciens* and *Agrobacterium vitis* strains. J Bacteriol 183:6852–6861

Bebber DP (2019) Climate change effects on Black Sigatoka disease of banana. Philos Trans R Soc B374:20180269. https://doi.org/10.1098/rstb.2018.0269

Bebber DP, Holmes T, Gurr SJ (2014) The global spread of crop pests and pathogens. Glob Ecol Biogeogr 23:1398–1407

Beddow JM, Pardey PG, Chai Y, Hurley TM, Kriticos DJ, Braun JC, Park RF, Cuddy WS, Yonow T (2015) Research investment implications of shifts in the global geography of wheat stripe rust. Nat Plant 1:15132

Bedos C, Cellier P, Calvet R, Barriuso E, Gabrielle B (2002) Mass transfer of pesticides into the atmosphere by volatilization from soils and plants: overview. Agronomie 22:21–33

Bregaglio S, Donatelli M, Confalonieri R (2013) Fungal infections of rice, wheat, and grape in Europe in 2030-2050. Agron Sustain Dev 33:767–776

Brown JKM, Hovmoller MS (2002) Aerial dispersal of pathogens on the global and continental scales and its impact on plant disease. Science 297:537–541

Burdon JJ, Thrall PH, Ericson AL (2006) The current and future dynamics of disease in plant communities. Annu Rev Phytopathol 44:19–39

Canto T, Aranda MA, Fereres A (2009) Climate change effects on physiology and population processes of hosts and vectors that influence the spread of hemipteran-borne plant viruses. Glob Chang Biol 15:1884–1894

Chakraborty S (2005) Potential impact of climate change on plant-pathogen interactions. Australas Plant Pathol 34:443–448

Chakraborty S, Datta S (2003) How will plant pathogens adapt to host plant resistance at elevated CO_2 under a changing climate? New Phytol 159:733–742

Chakraborty S, Pangga IB (2004) Plant disease and climate change. In: Gillings M, Holmes A (eds) Plant microbiology. BIOS Scientific Publishers, Abington, pp 163–180

Chen W, Wellings C, Chen X, Kang Z, Liu T (2014) Wheat stripe yellow rust caused by *Puccinia striiformis* f. sp. *tritici*. Mol Plant Pathol 15:433–446

Cheng C, Gao X, Feng B, Sheen J, Shan L, He P (2013) Plant immune response to pathogens differs with changing temperatures. Nat Commun 4:2530

Churchill ACL (2011) *Mycosphaerella fijiensis*, the black leaf streak pathogen of banana: progress towards understanding pathogen biology and detection, disease development, and the challenges of control. Mol Plant Pathol 12:307–328

Coakley SM, Scherm H, Chakraborty S (1999) Climate change and plant disease management. Annu Rev Phytopathol 37:399–426

Cohen SP, Liu H, Argueso CT, Pereira A, Vera Cruz C, Verdier V, Leach JE (2017) RNA-Seq analysis reveals insight into enhanced rice *Xa7*-mediated bacterial blight resistance at high temperature. PLoS One 12:e0187625

Crowl TA, Crist TO, Parmenter RR, Belovsky G, Lugo AE (2008) The spread of invasive species and infectious disease as drivers of ecosystem change. Front Ecol Environ 6:238–246

de Jong CF, Takken FLW, Cai X, de Wit PJGM, Joosten MHAJ (2002) Attenuation of *Cf*-mediated defense responses at elevated temperatures correlates with a decrease in elicitor-binding sites. Mol Plant-Microbe Interact 15:1040–1049

de Vallavieille-Pope C, Bahri B, Leconte M, Zurfluh O, Belaid Y, Maghrebi E, Huard F, Huber L, Launay M, Bancal MO (2018) Thermal generalist behavior of invasive *Puccinia striiformis* f. sp. *tritici* strains under current and future climate conditions. Plant Pathol 67(6):1307–1320. https://doi.org/10.1111/ppa.12840

De Wolf ED, Isard SA (2007) Disease cycle approach to plant disease prediction. Annu Rev Phytopathol 45:203–220

Dean R, Van Kan J, Pretorius ZA, Hammond-Kosack KE, Pietro A, Spanu PD, Rudd JJ, Dickman M, Kahmann R, Ellis J, Foster GD (2012) The top 10 fungal pathogens in molecular plant pathology. Mol Plant Pathol 13:414–430

Del Torro FJ, Aguilar E, Hernandez-Wallas FJ, Tenellado F, Chung BN, Canto T (2015) High temperature, high ambient CO_2, affect the interactions between three positive-sense RNA viruses and a compatible host differentially, but not their silencing suppression efficiencies. PLoS One 10:e0136062

Delcour I, Spanoghe, Uyttendaele M (2015) Impact of climate change on pesticide use. Food Res Int 68:7–15

Deutsch CA, Tewksbury JJ, Huey RB, Sheldon KS, Ghalambor CK, Haak DC, Martin PR (2008) Impacts of climate warming on terrestrial ectotherms across latitude. Proc Natl Acad Sci U S A 105:6668–6672

Dossa GS, Oliva R, Maiss E, Wydra K (2015) High temperature enhances the resistance of cultivated African rice, *Oryza glaberrima*, to bacterial blight. Plant Dis 100:380–387

Dyck P, Johnson R (1983) Temperature sensitivity of genes for resistance in wheat to *Puccinia recondite*. Can J Plant Pathol 5:229–234

Eastburn DM, Degennaro MM, Delucia EH, Dermody O, Mc Elrone AJ (2010) Elevated atmospheric carbon dioxide and ozone alter soybean diseases at SoyFACE. Glob Chang Biol 16:320–330

Eastburn DM, McElrone AJ, Bilgin DD (2011) Influence of atmospheric and climatic change on plant-pathogen interactions. Plant Pathol 60:54–69

Eizenberg H, Plakhine D, Hershenhorn J, Kleifeld Y, Rubin B (2003) Resistance to broomrape (*Orobanche* spp.) in sunflower (*Helianthus annuus* L.) is temperature dependent. J Exp Bot 54:1305–1311

Fischer EM, Knutti R (2015) Anthropogenic contribution to global occurrence of heavy-precipitation and high-temperature extremes. Nat Clim Chang 5:560–564

Fones HN, Gurr SJ (2017) Noxious gases and the unpredictability of emerging plant pathogens under climate change. BMC Biol 15:31–36

Garrett KA, Dendy SP, Frank EE, Rouse MN, Travers SE (2006) Climate change effects on plant disease: genomes to ecosystems. Annu Rev Phytopathol 44:489–509

Gautam HR, Bhardwaj ML, Kumar R (2013) Climate change and its impact on plant diseases. Curr Sci 105:1685–1691

Ghini R, Hamada E, Bettiol W (2008) Climate change and plant diseases. Sci Agric 65:98–107

Grulke NE (2011) The nexus of host and pathogen phenology: understanding the disease triangle with climate change. New Phytol 189:8–11

Gutierrez AP, Ponti L, d'Oultremont T, Ellis C (2008) Climate change effects on poikilotherm tritrophic interactions. Clim Chang 87:167–192

Hannukkala AO, Kaukoranta T, Lehtinen A, Rahkonen A (2007) Late-blight epidemics on potato in Finland, 1933-2002; increased and earlier occurrence of epidemics associated with climate change and lack of rotation. Plant Pathol 56:167–176

Hartleb H, Heitefuss R (1997) Abiotic and biotic influences on resistance of crop plants against fungal pathogens. In: Hartleb H, Heitefuss R, Hoppe HH (eds) Resistance of crop plants against fungi. Gustav Fischer Verlag, Jena, pp 298–326

Hasegawa H, Chatterjee A, Cui Y, Chatterjee AK (2005) Elevated temperature enhances virulence of *Erwinia carotovora* subsp. *carotovora* strain EC153 to plants and stimulates production of the quorum sensing signal, N-acyl homoserine lactone, and extracellular proteins. Appl Environ Microbiol 71:4655–4663

Helfer S (2014) Rust fungi and global change. New Phytol 201:770–780

Horino O, Mew TW, Yamada T (1982) The effect of temperature on the development of bacterial leaf blight on rice. Ann Phytopathol Soc Jpn 48:72–75

Hovmoller MS, Walter S, Bayles RA, Hubbard A, Flath K, Sommerfeldt N, Leconte M, Czembor P, Rodriguez-Algaba J, Thach T, Hansen JG, Lassen P, Justesen AF, Ali S, de Vallavieille-Pope C (2016) Replacement of the European wheat yellow rust population by new races from the centre of diversity in the near-Himalayan region. Plant Pathol 65:402–411

Huang L, Ren Q, Sun Y, Ye L, Cao H, Ge F (2012) Lower incidence and severity of tomato virus in elevated CO_2 is accompanied by modulated plant induced defence in tomato. Plant Biol 14:905–913

Hunjan MS, Singh PP, Lore JS (2014) Extracellular polysaccharides production underlining aggressiveness of rice bacterial blight pathogen *Xanthomonas oryzae* pv. *oryzae*. Indian Phytopathol 68:120–122

IPCC (2014) In: Field CB, Barros VR, Dokken DJ, Mach KJ, Mastrandrea MD, Bilir TE, Chatterjee M, Ebi KL, Estrada YO, Genova RC, Girma B, Kissel ES, Levy AN, MacCracken S, Mastrandrea PR, White LL (eds) Climate change 2014: impacts, adaptation, and vulnerability. Part A: global and sectoral aspects. I. Cambridge University Press, Cambridge/New York, p 1132

Jin S, Song YN, Deng WY, Gordon MP, Nester EW (1993) The regulatory VirA protein of *Agrobacterium tumefaciens* does not function at elevated temperatures. J Bacteriol 175:6830–6835

Jindal MM, Mohan C, Pannu PPS (2012) Status of stripe rust of wheat in Punjab during 2011–12 season. In: Proceedings of brain storming session. Department of Plant Pathology PAU, Ludhian, p 56

Jones RAC (2009) Plant virus emergence and evolution: origins, new encounter scenarios, factors driving emergence, effects of changing world conditions, and prospects for control. Virus Res 141:113–130

Jones RAC (2014) Virus disease problems facing potato industries worldwide: viruses found, climate change implications, rationalising virus strain nomenclature and addressing the potato virus Y issue. In: Navarre R, Pavek MJ (eds) The potato: botany, production and uses. CABI Publishing, Wallingford, pp 202–224

Jones RAC, Barbetti MJ (2012) Influence of climate change on plant disease infections and epidemics caused by viruses and bacteria. CAB Rev 7:1–32

Juroszek P, Tiedemann AV (2011) Potential strategies and future requirements for plant disease management under a changing climate. Plant Pathol 60:100–112

Keikotlhaile BM (2011) Influence of the processing factors on pesticide residues in fruits and vegetables and its application in consumer risk assessment. PhD dissertation, Ghent University, Ghent, Belgium

Kobayashi T, Ishiguro K, Nakajima T, Kim HY, Okada M, Kobayashi K (2006) Effects of elevated atmospheric CO_2 concentration on the infection of rice blast and sheath blight. Phytopathology 96:425–431

Kumar RS, Sakhtivel N (2001) Exopolysaccharide of *Xanthomonas* pathovar strains that infect rice and wheat crops. Appl Microbiol Biotechnol 55:782–786

Legler SE, Caffi T, Rossi V (2012) A nonlinear model for temperature-dependent development of *Erysiphe necator* chasmothecia on grapevine leaves. Plant Pathol 61:96–105

Loladze A, Druml T, Wellings CR (2014) Temperature adaptation in Australasian populations of *Puccinia striiformis* f. sp. *tritici*. Plant Pathol 633:572–580
Lyon B, Broders K (2017) Impact of climate change and race evolution on the epidemiology and ecology of stripe rust in central and eastern USA and Canada. Can J Plant Pathol 39:385–392
Malmstrom CM, Melcher U, Bosque-Pérezc NA (2011) The expanding field of plant virus ecology: historical foundations, knowledge gaps, and research directions. Virus Res 159:84–94
Mariette N, Androdias A, Mabon R, Corbière R, Marquer B, Montarry J, Andrivon D (2016a) Local adaptation to temperature in populations and clonal lineages of the Irish potato famine pathogen *Phytophthora infestans*. Ecol Evol 6:6320–6331
Mariette N, Mabon R, Corbière R, Boulard F, Glais I, Marquer B, Pasco C, Montarry J, Andrivon D (2016b) Phenotypic and genotypic changes in French populations of *Phytophthora infestans*: are invasive clones the most aggressive? Plant Pathol 65:577–586
Matros A, Amme S, Kettig B, Buck-Sorlin GH, Sonnewald U, Mock HP (2006) Growth at elevated CO_2 concentrations leads to modified profiles of secondary metabolites in tobacco cv. Samsun and to increased resistance against infection with Potato virus Y. Plant Cell Environ 29:126–137
Mboup M, Bahri B, Leconte M, de Vallavieille-Pope C, Kaltz O, Enjalbert J (2012) Genetic structure and local adaptation of European wheat yellow rust populations: the role of temperature-specific adaptation. Evol Appl 54:341–352
Medina MM (2017) Towards an inclusive risk-based approach in the cross-border ecommerce environment. 12th commission on phytosanitary measures for IPPC. Republic of Korea, pp 5–11
Milus EA, Seyran E, McNew R (2006) Aggressiveness of *Puccinia striiformis* f. sp *tritici* isolates in the South-Central United States. Plant Dis 90:847–852
Milus EA, Kristensen K, Hovmoller MS (2009) Evidence for increased aggressiveness in a recent widespread strain of *Puccinia striiformis* f. sp. *tritici* causing stripe rust of wheat. Phytopathology 99:89–94
Nancarrow N, Constable FE, Finlay KJ, Freeman AJ, Rodoni BC, Trebicki P, Vassiliadis S, Yen AL, Luck JE (2014) The effect of elevated temperature on Barley yellow dwarf virus-PAV in wheat. Virus Res 186:97–103
Nandal K, Sehrawat AR, Yadav AS, Vashishat RK, Boora KS (2005) High temperature-induced changes in exopolysaccharides, lipopolysaccharides and protein profile of heat-resistant mutants of *Rhizobium* sp. (*Cajanus*). Microbiol Res 160:367–373
Nazir N, Bilal S, Bhat KA, Shah TA, Badri ZA, Bhat FA, Wani TA, Mugal MN, Parveen S, Dorjey S (2018) Effect of climate change on plant diseases. Int J Curr Microbiol App Sci 7:250–256
Newbery F, Aiming Q, Bruce DLF (2016) Modelling impacts of climate change on arable crop diseases: progress, challenges and applications. Curr Opin Plant Biol 32:101–109
Ordonez N, Seidl MF, Waalwijk C, Drenth A, Kilian A, Thomma BPHJ, Ploetz RC, Kema GHJ (2015) Worse comes to worst: bananas and panama disease- when plant and pathogen clones meet. PLoS Pathog 11:e1005197
Paterson RRM, Lima N (2010) How will climate change affect mycotoxins in food? Food Res Int 43:1902–1914
Petzoldt C, Seaman A (2006) Climate change effects on insects and pathogens. In: Climate change and agriculture: promoting practical and profitable responses. New York State IPM Program, 630 W. North St., New York State Agricultural Extension Station, Geneva, pp 6–16
Prashar M, Bhardwaj SC, Jain SK, Datta D (2007) Pathotypic evolution in *Puccinia striiformis* in India during 1995–2004. Aust J Agric Res 58:602–604
Pritchard SG, Rogers HH, Prior SA, Peterson CM (1999) Elevated CO_2 and plant structure: a review. Glob Chang Biol 5:807–837
Samuel G (1931) Some experiments on inoculating methods with plant viruses, and on local lesions. Ann Appl Biol 18:494–507
Seguin J (2008) Human health in a changing climate: a Canadian assessment of vulnerabilities and adaptive capacity. Minister of Health in Canada, Ontario, p 524
Sharma-Poudyal D, Chen X, Rupp RA (2014) Potential oversummering and overwintering regions for the wheat stripe rust pathogen in the contiguous United States. Int J Biometeorol 58:987–997

Sparks AH, Forbes GA, Hijmans RJ, Garrett KA (2014) Climate change may have limited effect on global risk of potato late blight. Glob Chang Biol 20:3621–3631

Strand JF (2000) Some agrometeorological aspects of pest and disease management for the 21st century. Agric For Meteorol 103:73–82

Tang X, Cao X, Xu S, Jiang Y, Luo Y, Ma Z, Fan J, Zhou Y (2017) Effects of climate change on epidemics of powdery mildew in winter wheat in China. Plant Dis 101:1753–1760

Trebicki P, Nancarrow N, Cole E, Bosque-Perez NA, Constable F, Freeman AJ, Rodoni B, Yen AL, Luck JE, Fitzgerald GJ (2015) Virus disease in wheat predicted to increase with a changing climate. Glob Clim Chang 21:3511–3519

Trębicki P, Dáder B, Vassiliadis S, Fereres A (2017) Insect–plant–pathogen interactions as shaped by future climate: effects on biology, distribution, and implications for agriculture. Insect Sci 24:975–989

Uauy C, Brevis JC, Chen X, Khan I, Jackson L, Chicaiza O, Distelfeld A, Fahima T, Dubcovsky J (2005) High temperature adult plant (HTAP) stripe rust resistance gene *Yr36* from *Triticum turgidum* ssp. *dicoccoides* is closely linked to the grain protein content locus Gpc-B1. Theor Appl Genet 112:97–105

Vardharajula S, Ali SZ (2015) The production of exopolysaccharide by *Pseudomonas putida* GAP-P45 under various abiotic stress conditions and its role in soil aggregation. Microbiology 84:512–519

Vary Z, Mullins E, Mc Elwain C, Doohan FM (2015) The severity of wheat diseases increases when plants and pathogens are acclimatized to elevated carbon dioxide. Glob Chang Biol 21:2661–2669

Vera Cruz CM, Bai J, On I, Leung H, Nelson RJ, Mew TW, Leach JE (2000) Predicting durability of a disease resistance gene based on an assessment of the fitness loss and epidemiological consequences of avirulence gene mutation. Proc Natl Acad Sci U S A 97:13500–13505

Walter S, Ali S, Kemen E, Nazari K, Bahri BA, Enjalbert J, Hansen JG, Brown JKM, Sicheritz Pontén T, Jones J, de Vallavieille-Pope C, Justesen AF, Hovmoller MS (2016) Molecular markers for tracking the origin and worldwide distribution of invasive strains of *Puccinia striiformis*. Ecol Evol 6:2790–2804

Wang Y, Bao Z, Zhu Y, Hua J (2009) Analysis of temperature modulation of plant defense against biotrophic microbes. Mol Plant-Microbe Interact 22:498–506

Webb KM, Ona I, Bai J, Garrett KA, Mew T, Vera Cruz CM, Leach JE (2010) A benefit of high temperature: increased effectiveness of a rice bacterial blight disease resistance gene. New Phytol 185:568–576

Wingender J, Neu TR, Flemming HC (1999) What are bacterial extracellular polymeric substances? In: Wingender J, Neu TR, Flemming HC (eds) Microbial extracellular polymeric substances: characterization, structure and function. Springer, Berlin, pp 1–19

Wolfe DW, Ziska L, Petzoldt C, Seaman A, Chase L, Hayhoe K (2008) Projected change in climate thresholds in the Northeastern U.S.: implications for crops, pests, livestock, and farmers. Mitig Adapt Strateg Glob Chang 13:555–575

Wukasch R, Hofstra G (1977) Ozone and Botrytis interactions in onion-leaf dieback: open-top chamber studies. Phytopathology 77:1080–1084. https://doi.org/10.1094/Phyto-67-1080

Xiao S, Brown S, Patrick E, Brearley C, Turner JG (2003) Enhanced transcription of the *Arabidopsis* disease resistance genes *RPW8.1* and *RPW8.2* via a salicylic acid-dependent amplification circuit is required for hypersensitive cell death. Plant Cell 15:33–45

Zhang S, Li X, Sun Z, Shsao S, Hu L, Ye M, Zhou Y, Xia X, Yu J, Shia K (2015) Antagonism between phytohormone signaling underlies the variation in disease susceptibility of tomato plants under elevated CO_2. J Exp Bot 66:1951–1963

Zheng SJ, García-Bastidas FA, Li X, Zeng L, Bai T, Xu S, Yin K, Li H, Fu G, Yu Y, Yang L, Nguyen HC, Douangboupha B, Khaing AA, Drenth A, Seidl MF, Meijer HJG, Kema GHJ (2018) New geographical insights of the latest expansion of *Fusarium oxysporum* f.sp. *cubense* Tropical Race 4 into the Greater Mekong Subregion. Front Plant Sci 9:457. https://doi.org/10.3389/fpls.2018.00457

Potential for an Impact of Global Climate Change on Insect Herbivory in Cereal Crops

Finbarr G. Horgan

Introduction

Weather and climate affect multiple aspects of arthropod biology, behavior, and ecology (Cannon 1998; Bale et al. 2002). Despite the tremendous ecosystem services (e.g., pollination, decomposition, and pest regulation) provided by arthropods in agroecosystems, research into the possible implications of global climate change for crop production has focused predominantly on the disservices of insect herbivores (Cannon 1998; Rosenzweig et al. 2001; Hullé et al. 2010; Kocmánková et al. 2010; Svobodová et al. 2014; Battisti and Larsson 2015). A number of researchers have predicted increased pest incidences under future climate scenarios, and global climate change is frequently implicated as a cause of increasing damage to crop plants from pests (Hongyi et al. 2010; Hu et al. 2010b; Kocmánková et al. 2010; Hu et al. 2011; Lu et al. 2012; Hong et al. 2019). However, these ideas are beginning to be challenged by more holistic research and models (Zhao et al. 2016; Pecl et al. 2017; Sánchez-Bayo and Wyckhuys 2019; Tougeron et al. 2019; Tougeron and Tena 2019). The ecology of agroecosystems is complex, and the effects of even simple changes to crop management, such as adding fertilizer or deploying new crop varieties, are already difficult to predict (Horgan et al. 2016; Horgan et al. 2019a). Predicting the possible consequences of global climate change is therefore even more challenging, particularly since climate change occurs together with several other complex global changes (e.g., industrialization, globalization, and increasing data capture) (Horgan 2017a, 2017b; Sánchez-Bayo and Wyckhuys 2019).

This chapter addresses the implications of climate change for the complex interactions between plants and arthropods in agricultural systems. The chapter suggests that predictive models of the potential effects of climate change on crop pests should be developed using holistic frameworks that allow investigation of possible

F. G. Horgan (✉)
EcoLaVerna Integral Restoration Ecology, Bridestown, Kildinan, Co. Cork, Ireland
e-mail: f.horgan@ecolaverna.org

K. Jabran et al. (eds.), *Crop Protection Under Changing Climate*,
https://doi.org/10.1007/978-3-030-46111-9_5

alternative, and sometimes more credible, hypotheses. The chapter suggests that without careful consideration of the combined effects of compounding global changes – many of which are more easily amended than global climate – a changing climate could indeed accelerate the growth of pest populations and dramatically reduce production stability. Examples will be drawn from published studies of arthropod interactions with maize, wheat, and rice. These crops have been selected as three of the most important crops globally that together represent ~60% of global food consumption (Lobell et al. 2011). Furthermore, wheat is a predominantly temperate crop, whereas rice is predominantly tropical and subtropical (Fig. 1). Wheat and rice are also C3 crops with maize representing a C4 crop (Lobell et al. 2011). Each of these crops has an extensive associated literature related to key insect herbivores and other arthropods (Bosque-Pérez 1979; Hatchett et al. 1987; Ortega 1987; Heinrichs 1994). The chapter does not consider climate effects on decomposers or pollinators and the services they provide for agriculture; however, the following studies are good sources of information on potential climate effects on pollinators (Hegland et al. 2009; Byers 2017) and other beneficial arthropods (Menéndez et al. 2014; Coyle et al. 2017; Sánchez-Bayo and Wyckhuys 2019). The chapter highlights the need to acknowledge the ecological complexity of arthropod communities if we are to better develop future, climate-resilient production systems that avoid potential pest problems and maintain or increase agricultural productivity.

Trends in Global Climate Change

Changes in the global climate have included responses to fluctuations in solar irradiance and to decreased global volcanic activity, as well as feedbacks in the climate system that began in the mid-nineteenth century (Overpeck et al. 1997); however, changes in the current postglacial climate have accelerated in recent decades due to anthropogenic emissions of greenhouse gases (GHG) (Pachauri et al. 2014). Most of these emissions are directly related to the burning of fossil fuels (Herzog 2009); however, other factors, including deforestation, contribute ~20% of current global carbon dioxide (CO_2) emissions (Baccini et al. 2012). In recent years (since 2017), atmospheric CO_2 concentrations have surpassed 400 ppm. This is thought to be higher than at any time in the Earth's previous 800,000 years (Fig. 2). This CO_2 concentration continues to increase at a largely constant rate despite recent calls for reduced emissions (NASA 2019). The Intergovernmental Panel on Climate Change [IPCC] has selected four representative concentration pathways [RCPs] for climate modeling to predict future CO_2 concentrations (Pachauri et al. 2014). The first, RCP2.6, represents a scenario where global annual GHG emissions peak between 2010 and 2020 and decline thereafter. The RCP4.5 scenario represents a peak at around 2040 and the RCP6.0 a peak at around 2080. Based on the most probable RCPs, global CO_2 concentrations are expected to double before the end of this century [i.e., > 700 ppm under RCP6.0: (Pachauri et al. 2014)] (Fig. 2).

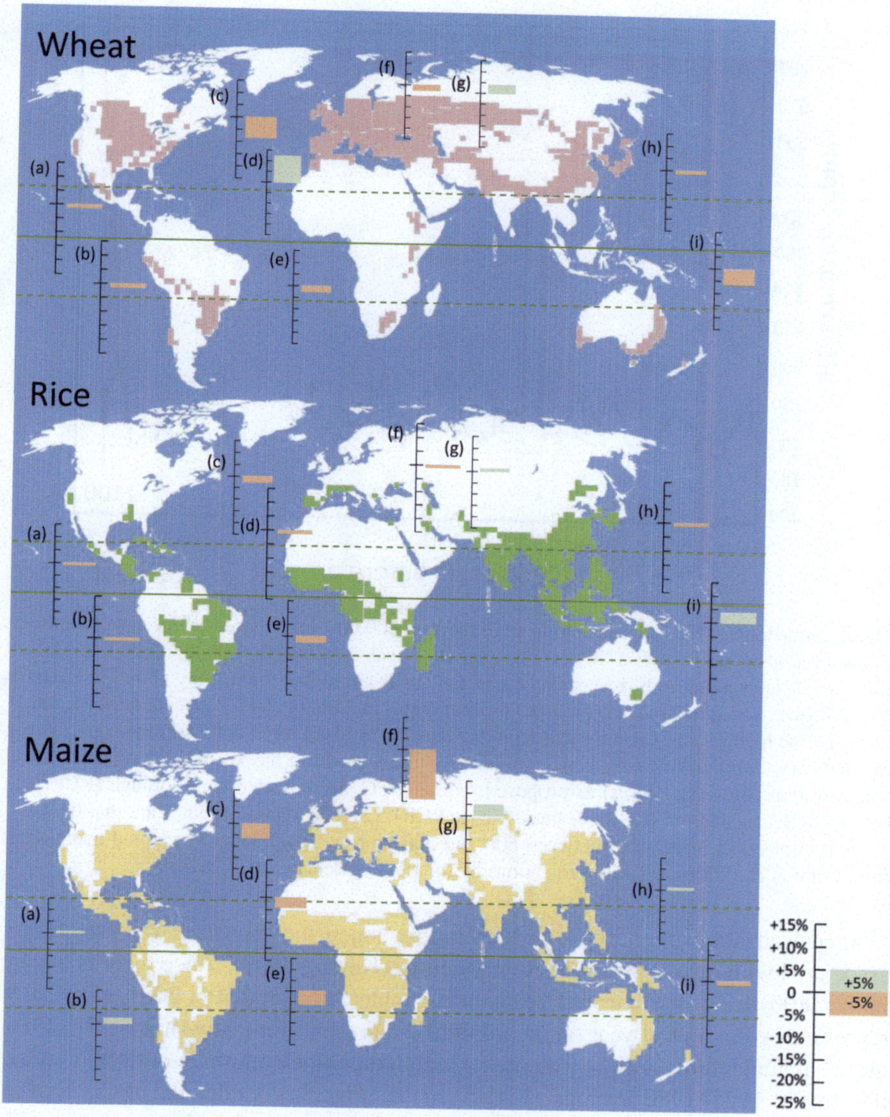

Fig. 1 The global distribution of wheat, rice, and maize redrawn with data from the FAO (2019). Estimated changes in production of each crop are indicated for (**a**) North and Central America; (**b**) the Caribbean and South America; (**c**) western and southern Europe; (**d**) eastern and northern Europe; (**e**) North Africa; (**f**) sub-Saharan Africa; (**g**) Central and Eastern Asia; (**h**) Western, Southern, and Southeast Asia; and (i) Oceania, using data presented by Ray et al. (2019) based on linear regression models using weather and crop data to estimate climate impacts from 1974 to 2008

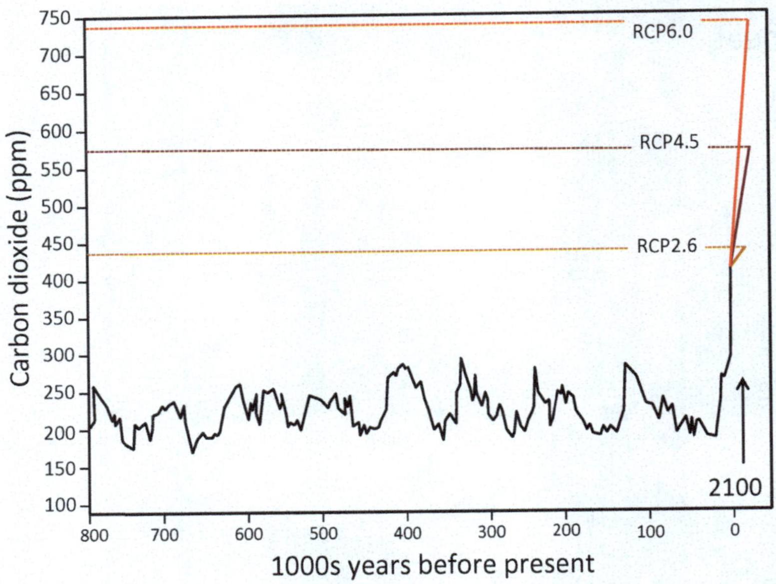

Fig. 2 Atmospheric CO_2 concentrations over the past 800,000 years based on ice core sampling. Concentrations of atmospheric CO_2 exhibit a series of peaks and troughs that represent interglacial and glacial (ice age) periods, respectively. Current levels are higher than at any period in the past 800,000 years and surpassed 400 parts per million (ppm) in January 2017. Current high concentrations are due to massive CO_2 emissions from the burning of fossil fuels, deforestation, agriculture, and industry. Estimates of CO_2 concentrations for the year 2100 based on three representative concentration pathways (RCPs) as proposed by the IPCC are indicated using orange (RCP2.6), brown (RCP4.5), and red (RCP6.0) lines. Under RCP4.5 and RCP6.0, estimates are that the CO_2 concentrations will increase to between 600 and 700 ppm by 2100. [Figure adapted using data from Lüthi et al. (2008) and with projections by the IPCC (Pachauri et al. 2014)]

Increased atmospheric CO_2 concentrations since the industrial revolution are associated with increasing global temperatures (Herzog 2009). Since records began in the 1800s, 18 of the warmest years on Earth have occurred after 2001, with 2016 representing the hottest year on record until the present (time of writing). The 2016 global annual temperature represented a high temperature anomaly of 0.98 °C relative to the 1951–1980 average (NASA 2019) (Fig. 3a). Under the IPCC RCPs, global warming is estimated to reach between 1.61 °C (RCP2.6) and 1.91 °C (RCP4.5) in 2046–2065 and between 1.61 °C and 2.81 °C (RCP6.0) in 2081–2100 (relative to the 1850–1900 reference period) (Pachauri et al. 2014). Increases in global temperatures have resulted in the retreat of glaciers and melting of polar ice to cause a rise in sea levels (Fig. 3b). It is estimated that globally, sea levels have risen by about 20 cm since the industrial revolution (Nicholls and Cazenave 2010; Pachauri et al. 2014). This has increased the frequency of flooding and saltwater intrusion in lowland coastal regions (Rozema and Flowers 2008; Hoeppe 2016). According to the RCP4.5 and RCP6.0 models, sea levels are expected to rise by

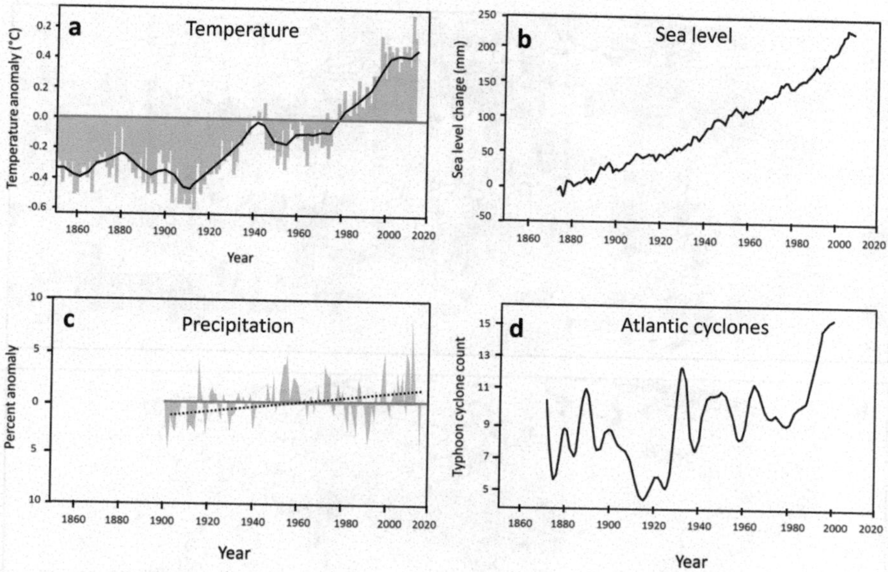

Fig. 3 (**a**) Global temperature anomalies redrawn using data from Brohan et al. (2006) with recent temperatures added according to NASA (2019). (**b**) Changes in sea level redrawn using data from NASA (2019). (**c**) Anomalies in global precipitation redrawn according to the EPA (2019). (**d**) The frequency of Atlantic hurricane activity redrawn according to Mann and Emanuel (2006)

between 0.40 m and 0.48 m before 2100 [from base levels during the period 1986–2005] (Pachauri et al. 2014).

Historical records also indicate a gradual increase in global precipitation since the early 1900s, with the wettest years largely coinciding with periods of sustained high temperature (1940–1950 and 2000–2020) (Fig. 3c) (EPA 2019). However, this increase in precipitation has not been experienced uniformly over the planet but, rather, has been concentrated in temperate regions. For example, rainfall in the Sahel of northern Africa has declined dramatically since the 1970s (JISAO 2019). The distribution of rainfall largely follows changes in global cloud cover, with trends reflecting a poleward retreat of midlatitude storm tracks and an expansion of subtropical dry zones (Norris et al. 2016) (Fig. 4a). High temperatures and a decrease in rainfall in arid regions have led to the expansion of dryland areas. Huang et al. (2015) estimate that >50% of the total land surface will be covered in drylands by 2100. This will lead to a reduction in carbon sequestration that will further accelerate regional warming.

Changes in global climate include complex positive feedback systems, including albedo effects due to snow melt and to changes in vegetation cover (Gildor and Tziperman 2003; Gibbard et al. 2005) (Fig. 5). Changes in climate also interact with land cover change to exacerbate the impacts of habitat loss, which magnifies weather effects on abiotic and biotic components of the environment at local or regional scales (Mantyka-Pringle et al. 2015). For example, hot, dry conditions together with deforestation and the clearing of scrublands increase atmospheric dust loads. It is

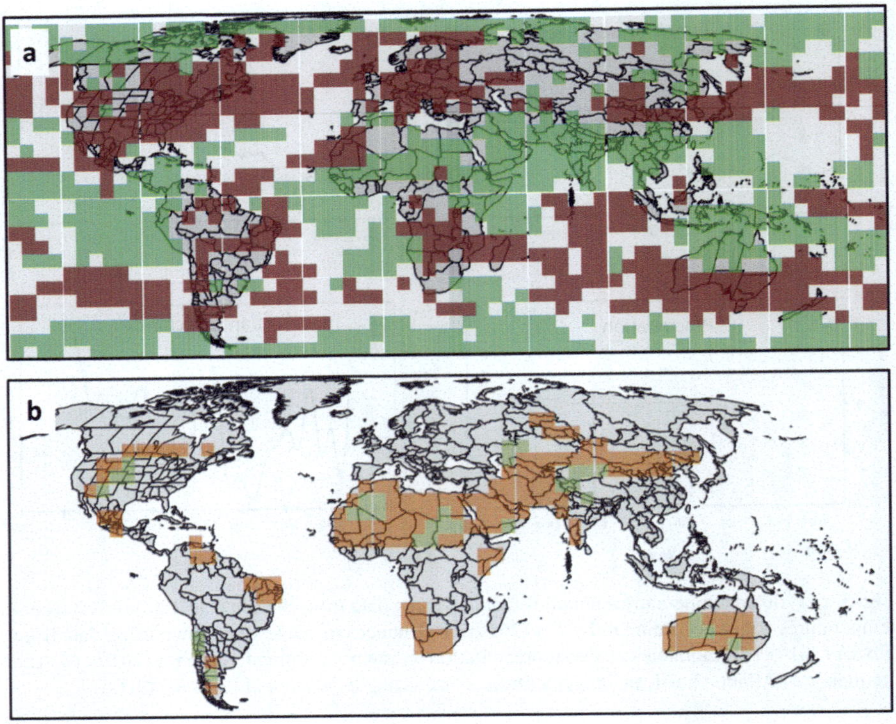

Fig. 4 (**a**) Cloud trend patterns from 1983 to 2009 redrawn and simplified from a simulation model presented by Norris et al. (2016) reflecting a polar migration of clouds (red) and reduced cloud cover (green) over the expanding subtropical dry zones. (**b**) Estimated areas of increasing (brown) and decreasing (green) dust flux (g/m^2/year) including natural and cultivated soils between the periods 1970–1980 and 2070–2080. The image (**b**) is simplified from a simulation model presented by Tegen et al. (2004). See original papers for full details related to changes in clouds and dust

estimated that over half of the current estimated dust loads originate from anthropogenically disturbed soils and > 10% is generated by agricultural activities (Tegen et al. 2004). Climate change and land-use change will further alter dust emissions with decreases predicted in some areas but increases, largely influenced by cultivated soils, in others (Tegen et al. 2004) (Fig. 4b). These dusts will be transported by future winds.

Changes in sea temperatures have influenced the strength and direction of winds and the occurrence of typhoons and hurricanes (Mann and Emanuel 2006; Knutson et al. 2015; Takagi and Esteban 2016). It is estimated that global wind speeds have decreased by 5–15% over the last 30 years and are expected to continue to decrease into the future (Barton 2014). However, whereas average wind speeds have decreased, the frequency and intensity of storms have increased in some areas (Knutson et al. 2015). For example, the number of tropical cyclones making landfall around Leyte Island in the Philippines has been steadily increasing since the begin-

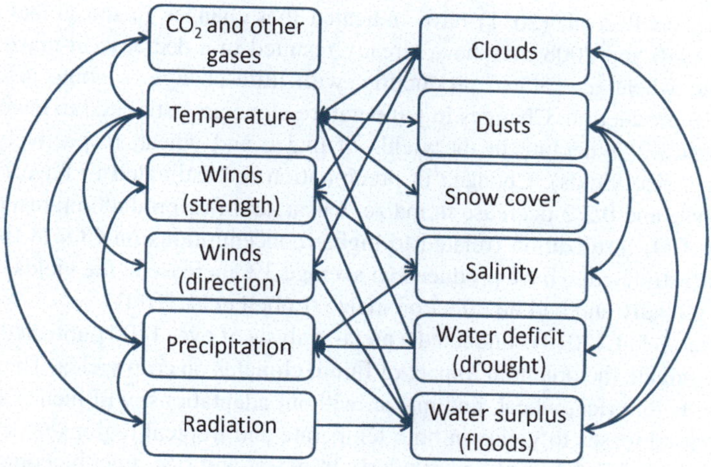

Fig. 5 Whereas much of the literature on climate effects on crops has focused on increasing global temperatures and increasing CO_2 concentrations, other related effects such as the concentration of particulate matter in the air (dusts), or changes in wind speed and direction, can have significant effects on crop plants and affect the nature of plant-herbivore interactions and community interactions at higher trophic levels. Global climate change includes a range of global effects (indicated on the left) that produce other more localized changes in agricultural environments (indicated on the right). These are influenced by a range of positive feedback systems (indicated by curved arrows) that further accelerate changes

ning of the millennium (Takagi and Esteban 2016), and there has been a noted increase in the incidence of Atlantic hurricanes that appears largely correlated with temperature anomalies in recent decades (Mann and Emanuel 2006) (Fig. 2d). Apart from the increasing intensity of storms, there have also been shifts in rainfall patterns and in the trajectories of storm tracks and consequently in the relative vulnerabilities of different coastal areas to storms and storm surges (Knutson et al. 2015; Loo et al. 2015).

Together these trends depict global changes that are not linear, but a combination of gradual changes together with an increasing frequency and magnitude of extreme weather events. Harris et al. (2018) refer to this as a press-pulse system and suggest that to be more realistic, predictive models should be based on press-pulse frameworks.

The Impacts of Climate Change on Crop Yield Potential

Based primarily on crop physiology, a number of studies have modeled the potential effects of recent and predicted future climate conditions on crop productivity. Several of these models indicate a general decrease in the global production of key food crops, including wheat, maize, and rice, as well as increasing temporal variability in crop yields (Lobell et al. 2011; Challinor et al. 2014; Ray et al. 2019). For example, Lobell et al. (2011) have indicated that changes in the global climate between 1980 and 2008 may have already resulted in a decrease in maize (about 3.8%) and wheat (ca. 5.5%) production, with little change or some increase in global rice production. Changes in temperature alone are estimated to have caused a 3.1% and 4.9% decline in the yields of maize and wheat, respectively (0.1% increase in rice yields). Changes in precipitation are estimated to have caused a 0.7%, 0.6%, and 0.2% decrease in maize, wheat, and rice production, respectively. However, CO_2 fertilization (related to higher concentrations on CO_2 in the atmosphere) is estimated to have produced an average 3% increase in the yields of wheat and rice globally and had no effect on maize (Lobell et al. 2011).

Challinor et al. (2014) conducted a meta-analysis of over 1700 published simulations to evaluate the potential impact of future climates on crop yields. Their analysis suggests that rice, wheat, and maize, without adaptations, will incur variability and increased losses to yields in both temperate and tropical regions under a 2 °C warming. Some of the losses, particularly in wheat and rice, may be countered by adaptive management (that increased simulated yields by 7–15%). Based on the projected variability and declines in future crop yields, pest management is likely to play an increasingly prominent role in future farm operations regardless of whether crop damage from insect pests increases or not. This is because farmers will often attribute to insects and diseases variability in yields that is actually due to weather or poor crop management (Horgan 2017b). Such farmers will mistakenly respond to variable yields by increasing their investments in pest management actions.

Climate Change and Arthropod Herbivory

Arthropods are predominantly ectotherms that regulate their body temperatures using environmental heat sources (Bale et al. 2002; Bale and Hayward 2010). As such, the behavior and ecology of arthropods are heavily influenced by ambient temperatures. Other aspects of weather and climate that influence arthropod behavior include precipitation and relative humidity, winds and thermal drafts, and solar radiation or light intensity, including photoperiods (Rosenberg and Magor 1987; Eizaguirre et al. 1994; Barton 2014; Wu et al. 2018). Researchers have explored the influence of weather on the physiology, behavior, and population dynamics of synanthropic arthropods for several decades. A large number of case studies have correlated weather data with insect populations or crop damage (Mochida et al. 1987; Isichaikul and Ichikawa 1993; Fang et al. 2013; Ali et al. 2019) and herbivore outbreaks (Hu et al., 2010a; Chen et al. 2013; WMO/FAO 2016). For example, studies have indicated that population fluctuations of aphids are correlated with temperatures in the years previous to field observations (Brabec et al. 2014). Ambient temperatures have been correlated with populations of skippers, *Pelopidas mathias* (Patel et al. 2011), and leaf folders, *Cnaphalocrocis medinalis*, in rice (Kaushik 2011; Patel et al. 2011). High springtime temperatures associated with El Niño/Southern Oscillation (ENSO) events have been linked to outbreaks of forest caterpillars (Horgan et al. 1999; Myers 2000; Horgan 2005) and rice planthoppers (Homoptera: Delphacidae) (Morishita 1992; Susanti et al. 2010). Outbreaks of migratory insects, such as the desert locust, *Schistocerca gregaria*, are associated with wind patterns and storms (WMO/FAO 2016). Observations such as these cannot be attributed to global climate change, although they do indicate the potential for a changing climate to impact arthropod populations. Historical records indicate that weather-related outbreaks of insects have predated large-scale anthropogenic increases in CO_2 and other greenhouse gases [e.g., planthoppers in Japan (Dyck and Thomas 1979) and locusts in Africa (WMO/FAO 2016)]. Furthermore, just as high temperatures are associated with high densities and outbreaks of pests, adverse weather – including floods and drought, heavy rains, high winds, or unseasonably cold spells – can also cause declines in herbivore populations (Kaushik 2011; Patel et al. 2011; Zhang et al. 2013; Lacoste et al. 2015).

Because of an increasing emphasis on global climate change, researchers should be vigilant against confirmation bias when investigating the relation between climate and recent changes in pest populations. Some criteria for attributing pest populations to climate change are therefore necessary. In recent decades, data has begun to accumulate on the large-scale and long-term dynamics of pest insects; therefore, it is now possible to examine correlations between weather parameters and insect populations that encompass several years of directional change in weather (Hullé et al. 2010; Ali et al. 2014, 2019; Brabec et al. 2014). In most cases, such correlation techniques (with their limitations due to cause-and-effect issues) can only be used to generate hypotheses and not to test them. Increasing the temporal or spatial scales of research and incorporating an increasing number of factors to address possible alternative hypotheses will strengthen the inferential power of statistical models. Indeed, if climate change is a result of anthropogenic activities, then multiple parallel changes are to be expected as a general rule (e.g., climate and landscape change, climate and industrial output, or climate and global travel). The most realistic models of the possible effects of climate change on insect populations will, therefore, necessarily be the most complex. A perusal of current literature on climate change indicates that models are, indeed, becoming more holistic and have begun to incorporate other factors together with weather data (Hu et al. 2010a; Zhao et al. 2016). Furthermore, an increasing number of research papers report the results of manipulative experiments, and there is a tendency for researchers to include increasing numbers of independent factors (e.g., temperature and humidity, temperature and atmospheric CO_2 concentration, or temperature and natural enemies) in their experimental designs (Kim et al. 2001; Fantinou et al. 2003; Fang et al. 2013; Shi et al. 2014; Xu et al. 2019). Nevertheless, in terms of predicting future pest scenarios, the state of the art is still at its infancy, and there is ample space to improve on current knowledge.

The following sections will examine evidence for the potential effects of climate change on arthropod herbivores (mainly insects) bearing in mind the many limitations inherent to this type of research as outlined above. The potential effects of climate change can be divided into four main categories: (1) direct effects of climate on arthropod herbivores, (2) indirect plant-mediated climate effects on arthropod herbivores, (3) indirect biodiversity-mediated climate effects on arthropod herbivores, and (4) indirect crop management-mediated climate effects on arthropod herbivores. These categories are not exclusive. For example, crop management will have large effects on natural enemies (part of the biodiversity-mediated component) and also influence plant-mediated effects. To provide information relative to the large assemblages of arthropod herbivores affecting crops and the wide range of local regional climates under which crops are attacked, each section will also consider four key herbivore attributes that determine broad categories of responses to climate. These key attributes are (1) whether the herbivore population is predominantly tropical or temperate, (2) whether the population is migratory or nonmigratory, (3) whether the species is specialist or generalist, and (4) whether populations occur in the species' native or introduced range.

Direct Effects of Climate on Arthropod Herbivores

Temperature is the principal climatic factor directly influencing arthropod herbivores. As outlined in a review by Bale et al. (2002), CO_2 and radiation do not directly affect arthropods and evidence for the direct effects of precipitation is scarce. Much of the current knowledge of temperature effects on insect herbivores has been derived from measurative experiments conducted in climate chambers. However, the application of such information to climate change models is limited. For example, arthropod species have greater genetic variability and genotypic flexibility than can be observed from laboratory experiments based on small, often laboratory-reared populations (Krysan et al. 1984; Levine et al. 1992a; Levine et al. 1992b). Arthropods also display varying degrees of phenotypic plasticity in their responses to temperature (Lu et al. 2016), and many species can acclimatize to temperature extremes through gradual exposure to suboptimal conditions during development (Gillyboeuf et al. 1994; Ma and Ma 2012; Piyaphongkul et al. 2014). These mechanisms of resilience to climate change are inherent to arthropods but often overlooked in modeling studies. Finally, most arthropod herbivores are mutualists that require a variety of endosymbiotic microorganisms, including bacteria and yeasts, in order to survive (Hansen and Moran 2014; Ferrater et al. 2015; Horgan et al. 2019b). Evidence suggests that endosymbionts can sometimes determine herbivore responses to temperature (Chen et al. 2009; Horgan and Ferrater 2017). For example, in a study by Chen et al. (2009), the primary endosymbiont *Buchnera aphidicola* was sensitive to high temperatures (≥ 35 °C), but its aphid host, *Aphis craccivora*, was most sensitive to low temperatures (≤ 10 °C). A recent study has shown that two bacterial symbionts in the aphid *Acyrthosiphon pisum* confer tolerance to high temperatures, whereas a third bacterium was a liability at the same temperatures (Russell and Moran 2005). Endosymbionts could therefore also promote herbivore resilience to a changing climate and may undergo more rapid directional selection for heat tolerance than their insect hosts.

Climate and Arthropod Range Expansion

Changes in climate, particularly increasing winter and springtime temperatures, will create conditions that are more favorable for tropical, subtropical, and warm temperate species to increase their ranges poleward (Hu et al. 2015; Forrest 2016; Tougeron et al. 2019). Considerable evidence of the poleward range expansion of herbivores and other arthropod functional groups has been gathered over recent decades (Tougou et al. 2009; Forrest 2016; Pecl et al. 2017). Much of this evidence is based on species with relatively restricted distributions. In some cases, such narrow range species have become less common at lower latitudes presumably due to their limited tolerance of high temperatures or low humidity (Franco et al. 2006). Because many of the most important crop pests are species with wide distribution ranges and many are not native in their current distribution ranges, observations of poleward extensions in the distribution of crop pests are rare. Only two cases have been documented in recent years. The range of the southern green shield bug, *Nezara viridula*, had apparently shifted 85 km northward in Japan between the 1960s and 2008, probably because of more favorable overwintering conditions. The species also declined in parts of its southern range (Tougou et al. 2009; Yukawa et al. 2009) (Fig. 6a). In China, there has been a northerly shift in outbreaks of the orange blossom wheat midge, *Sitodiplosis mosellana*. Outbreaks and damage to wheat shifted northward at a rate of 58.8 km per decade between the 1950s and 2010s. Furthermore, phenological matching between adult emergence and wheat heading in the new range has apparently increased midge-related losses to wheat production (Wu et al. 2019) (Fig. 6b).

Crops may also be attacked by a greater diversity of herbivores as the distributions of several different species shift poleward. For example, the number of aphid species identified at eight sites ranging from Scotland to southern France (extending

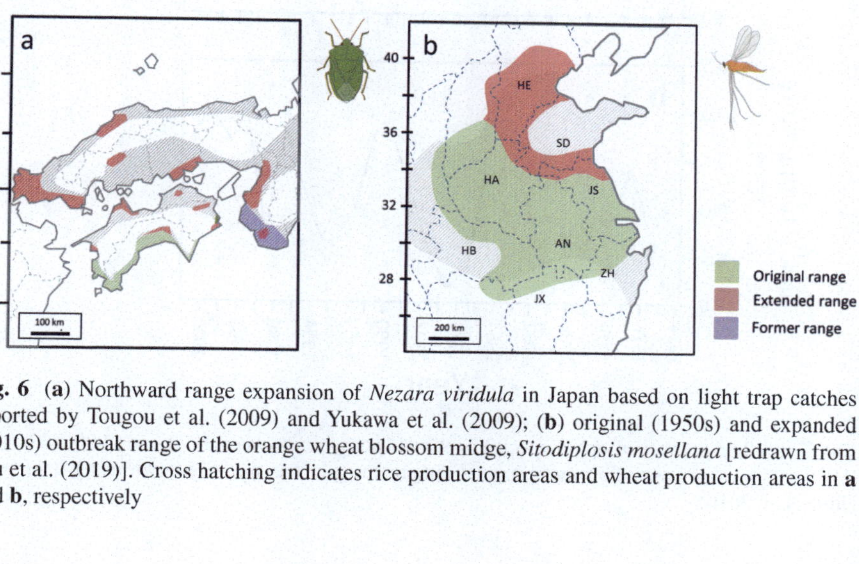

Fig. 6 (**a**) Northward range expansion of *Nezara viridula* in Japan based on light trap catches reported by Tougou et al. (2009) and Yukawa et al. (2009); (**b**) original (1950s) and expanded (2010s) outbreak range of the orange wheat blossom midge, *Sitodiplosis mosellana* [redrawn from Wu et al. (2019)]. Cross hatching indicates rice production areas and wheat production areas in **a** and **b**, respectively

over 2000 km) of the EXAMINE [aphid observation] network had increased by over 20% in the 30 years prior to 2010. The increase in diversity followed local temperature increases over the same period. This represented an increase of 15 species for every °C increase in temperature (Hullé et al. 2010) (Fig. 7). Whether this will translate into increased damage to crops and consequent yield losses is not clear. Agroecosystems typically include a diversity of arthropod herbivores, but only a few species will directly affect yields. Indeed, evidence suggests that a greater diversity of herbivores can lead to greater stability of arthropod communities and potentially reduce pest outbreaks (Ings et al. 2009; Horgan et al. 2019a). Furthermore, herbivores represent only a portion of the arthropod community that has increased its distribution range. As will be discussed in Sect. 7, climate-related changes in distribution ranges of the natural enemies of crop pests will also reduce the risks of increased herbivore damage. In the same way that arthropods respond to warmer temperatures by increasing their ranges poleward, many species are predicted to expand their ranges to higher elevations. For example, the spotted stalk borer, *Chilo partellus*, has spread rapidly in Africa since its introduction in the 1930s, in some regions displacing the endemic maize stalk borer, *Busseola fusca* (Kfir 1997). Predictive models based on life-stage responses to elevated temperatures indicate that the spotted stalk borer could further expand its distribution to maize-intensive farmlands in the highland tropics of East Africa (Khadioli et al. 2014).

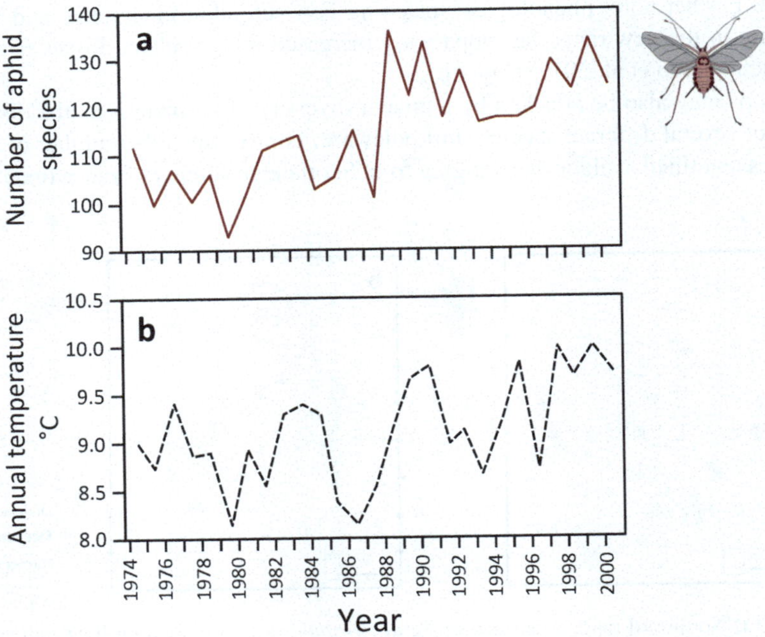

Fig. 7 (**a**) The number of aphid species captured each year between 1974 and 2000 at Preston (UK) with (**b**) corresponding mean annual temperatures at the site. (Redrawn using data from Hullé et al. 2010)

Species range expansion either poleward or to higher elevations will occur as a progressive establishment of populations over several generations. Therefore, as climate change occurs, arthropod monitoring programs such as the EXAMINE network will record species at successive stages of range expansion, including sterile invasions, temporary isolated populations that may eventually become extinct, or established populations that are disjunct from the original continuous distribution (Gorodkov 1986) (Fig. 8). As global temperatures increase, some species may also undergo range reductions at lower latitudes or lower altitudes. Tropical and subtropical insects in particular may be less tolerant of predicted climate changes than temperate species. For example, using temperature profiles and simulation models, crop damage from the fall armyworm, *Spodoptera frugiperda*, is predicted to decline in the southern USA and much of South America (Ramirez-Cabral et al. 2017), and brown planthopper, *Nilaparvata lugens*, and pink stem borer, *Sesamia inferens*, populations are predicted to decline in some parts of India as a result of reduced fecundity and lower survival under higher temperatures (Sujithra and Chander 2013; Krishnan and Chander 2015). Furthermore, many tropical montane species with narrow distribution ranges will be vulnerable to extinction where opportunities for redistribution from high elevations at low latitudes to similar biomes at higher latitudes are restricted by inherent physiological limits (Janzen 1967).

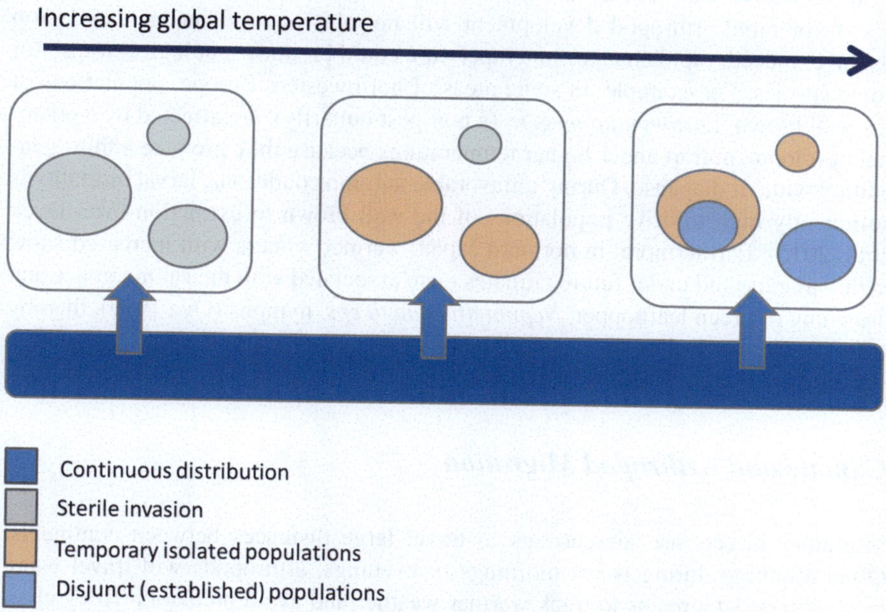

Fig. 8 Successive stages of invasion and establishment of species during range expansion. (Adapted from Gorodkov 1986)

Climate and Voltinism

A warmer climate can affect populations of multivoltine species by increasing the number of generations they complete each cropping season. For example, predicted higher temperatures of 2 °C in northern Europe are expected to increase the number of generations for some aphid species by as many as five per year (e.g., from 18 to 23 in the UK). This is partly because many aphids in northern Europe presently occur under suboptimal temperature conditions, because aphid populations can survive and remain active all year and because higher temperatures favor parthenogenesis (Hullé et al. 2010). Based on climate models, brown planthopper voltinism is predicted to increase by <0.5, 0.5–1.0, and 1.0–1.4 generations in the period between 2021 and 2050 in northern, central, and southern China, respectively (Hu et al. 2015). Other factors that will increase arthropod voltinism under global warming include earlier migration, earlier egg hatch or an earlier break in diapause after winter, as well as increased development rates. Already, ~ 95% of aphid species from the EXAMINE network have displayed increasingly early spring migrations since records began in the 1970s (Hullé et al. 2010).

Studies of crop pests in Europe suggest that as temperatures continue to rise, important pests, including the European corn borer, *Ostrinia nubilalis*, and the aphids *Rhopalosiphum padi* and *Sitobion avenae*, will increase the number of generations they complete in northern regions but will complete fewer generations in southern regions as ambient humidity declines (Kocmánková et al. 2010; Brabec et al. 2014, Svobodová et al. 2014). Furthermore, gradual increases in temperature that favor rapid arthropod development will not always result in increased crop damage; indeed, rapid changes in temperature could produce “ecological traps” for some species. For example, in some areas of northwestern Europe, populations of the wall brown, *Lasiommata megera* (a non-pest butterfly), are affected by a potential development trap under higher temperatures because they produce a third generation without diapause. During unfavorable autumn conditions, larval mortality is sufficiently high to drive populations of the wall brown to extinction (Van Dyck et al. 2015). Furthermore, in northern Japan, warmer winters with increased snow melt – as expected under future climates – are associated with the submergence and drowning of green leafhopper, *Nephotettix cincticeps*, nymphs (Ôya 1979), thereby reducing the survival of late season generations.

Climate and Arthropod Migration

Migratory insects use air currents to travel large distances between continents. Often alighting during warm mornings or evenings, arthropods will travel using high-altitude jet streams to track warmer weather and avoid predators (Riley et al. 1994, 1995; Kisimoto and Sogawa 1995; Sedlock et al. 2019). This strategy allows some insects to benefit from constant temperatures and new plant growth during tropical wet seasons and from emerging vegetation as temperatures increase and photoperiods extend during the subtropical and temperate spring or summer. Migration plays an important role in determining pest population traits such as virulence against resistant crop varieties and insecticide resistance (Matsumura et al. 2008; Horgan et al. 2017a). A number of important crop pests undergo such migrations. Among the best studied are populations of brown and white-backed (*Sogatella furcifera*) planthoppers that feed on rice in East Asia (Riley et al. 1994; Kisimoto and Sogawa 1995) (Fig. 9).

The key role of wind transport in planthopper migration has been highlighted through a number of studies that traced migration routes by backtracking wind conditions, landing dates, and insect development stages (Otuka et al. 2008; Otuka et al. 2012). Because planthoppers converge during cyclones and are deposited during storms and rain (Kisimoto and Sogawa 1995), a changing climate could have significant effects on migrating populations. For example, climate models depict directional changes in wind patterns that have already been noted in the South China Sea (Wang et al. 2009a; Loo et al. 2015), an important region for planthopper migration (Fig. 9). Furthermore, an increase in the number and strength of typhoons in the

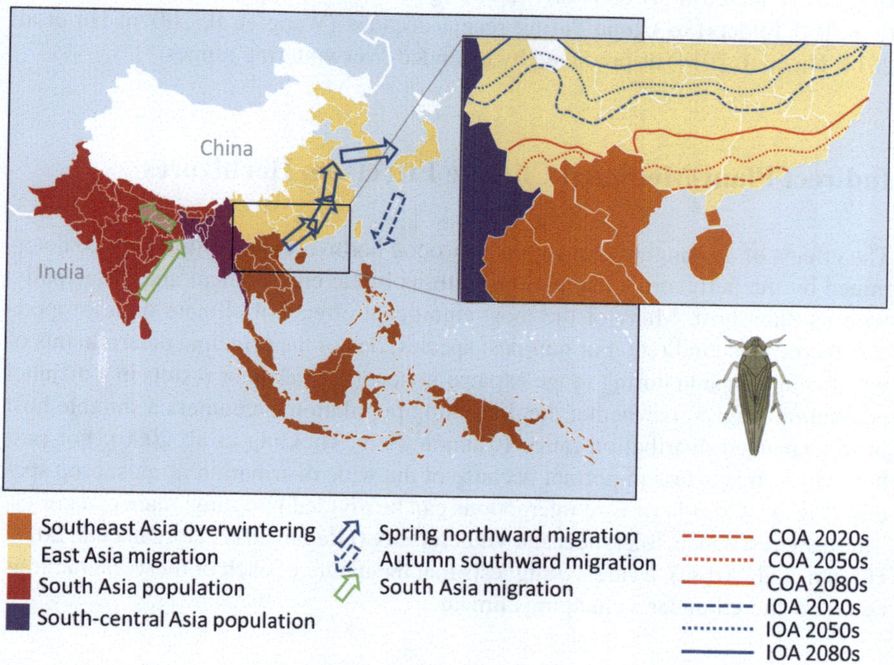

Fig. 9 Populations of brown planthopper, *Nilaparvata lugens*, in South, Southeast, and East Asia based on evidence from trapping programs and virulence data reported from a range of studies (Wada et al. 1987; Riley et al. 1994; Kisimoto and Sogawa 1995; Horgan et al. 2017a) (arrows indicate successive generations of migration). The inset map presents the estimated northward extension of the constant overwintering area (COA ≥ 16 °C isocline) and intermittent overwintering area (IOA ≥ 10 °C isocline) based on future climate scenarios (Hu et al. 2015)

north Pacific (Wang et al. 2009a; Takagi and Esteban 2016) could result in more frequent outbreaks of planthoppers and other migratory pests because of converging winds and a more rapid displacement of populations.

A further factor that could affect planthopper migrations is a potential northward extension of the species' overwintering boundaries (Fig. 9). Planthoppers are incapable of overwintering at temperatures below ~10–16 °C. Using data from 289 meteorological stations and published temperature thresholds for overwintering survival, recent models have predicted that between 1951 and 2010, the overwintering range of the brown planthopper has extended by 2–3 degrees of latitude (about 90 km) in some regions and that the northward progression of the overwintering boundary (represented by the 10 °C isocline) has accelerated in recent decades (Lu et al. 2012). In a similar study aimed at predicting the future limits of the overwintering boundary using standard IPCC climate prediction models, the intermittent overwintering area (represented by the ≥10 °C isocline) was predicted to increase by 11, 24, and 44% and the constant overwintering area (represented by the ≥16 °C isocline) by 66, 206, and 477% during the 2020s, 2050s, and 2080s, respectively (Hu et al. 2015). There is still little direct evidence of extended planthopper overwintering ranges in China (Luo et al. 2013); however, evidence from light trapping suggests an increasingly early arrival of migratory planthoppers and other rice pests (e.g., leaf folders) to China during recent decades (Wang et al. 2009b; Hu et al. 2011; Hu et al. 2014) as predicted by extended overwintering ranges.

Indirect Plant-Mediated Climate Effects on Herbivores

The effects of a changing climate on arthropod herbivores will ultimately be determined by the herbivore's interactions with its biotic environment and, principally, with its plant host. Many of the most significant effects of climate on arthropods will therefore be indirect. For non-pest species, one of the principal determinants of whether an invasion during range expansion becomes sterile or results in a disjunct population (Fig. 8) is whether the dispersing population encounters a suitable host in its expanded distribution range (Cannon 1998; Hickling et al. 2006). For pest herbivores, this is less important because of the wide distribution of most crop species (Fig. 1). Crop-herbivore interactions can be divided into three main categories. These are resistance, tolerance, and vulnerability (Horgan 2012; Horgan et al. 2016; Horgan et al. 2018c). Evidence suggests that the nature of each of these interactions could be altered under a changing climate.

Effects of Changing Climate on Host-Plant Resistance

Resistance is a plant's ability to defend against attacking herbivores and reduce damage. Resistance is normally divided into antixenotic components that deter insects from ovipositing or initiating feeding on the host and antibiosis components that reduce the fitness of herbivores feeding on the host (Horgan 2012). These components are determined by the physical and biochemical properties of the plant, including the presence/absence of trichomes and surface waxes, the composition of volatiles emitted from the plant, the presence/absence of toxins or antifeedants, and other factors (Salminen et al. 2005; Pelletier et al. 2013; Xu et al. 2019). Because resistance is a relative term, the effects of a changing climate on resistance can only be assessed through comparative experiments. Research on the potential impacts of climate change on resistance has mainly focused on the effects of increasing (high) temperatures, CO_2, dusts, and salt on the ability of crop plants to defend against herbivores. Condition-dependent changes in resistance are best detected in experiments as significant interactions between "climatic conditions" and "host plant" where physical condition (optimal and extreme) and host plant (susceptible and resistant) have at least two levels. However, very few studies have taken this approach. Instead, most studies have assessed whether changing conditions affect herbivore fitness on a single host genotype. These studies have indicated that changes in temperature may increase, decrease, or have no impact on the relative strength of resistance depending on the host species and the arthropod species involved. For example, in studies with tomatoes and sorghum, resistance to various insect herbivores declined under low temperatures (Schweissing and Wilde 1979; Stamp and Osier 1998). Cold weather has also been associated with a decline in the resistance of flowering wheat to the aphid *Rhopalosiphum padi* because of a reduction in the levels of potassium and copper in flowers during frost events (Lacoste et al. 2015). Wang et al. (2010) suggested that the rice variety IR36 had an optimal temperature for resistance at about 25 °C. The authors largely associated declining resistance at higher temperatures (above 25 °C) with increases in the content of soluble sugars in the plant tissues. The effects of temperature on resistance can also be the result of complex interactions between insects, plants, and other elements of their biotic environment. For example, resistance of tall fescue and perennial ryegrass against the fall armyworm, *Spodoptera frugiperda*, increases as plant growth temperatures increase from 15 to 25 °C. Higher anti-herbivore resistance was due to increases in the concentrations of alkaloids produced by the plants and by endophytes located in the plant roots. In the absence of endophytes, resistance declined (Salminen et al. 2005).

Understanding the effects of atmospheric CO_2 concentrations on host-plant resistance is complicated because CO_2 affects the nutritional quality of the host plant and therefore, although it may reduce herbivore fitness, it can also increase the relative damage caused by herbivores at high CO_2 concentrations. For example, in a study by Li et al. (2013), elevated CO_2 reduced the fitness of leaf folders on rice. This was associated with an increase in food consumption but a decline in the conversion

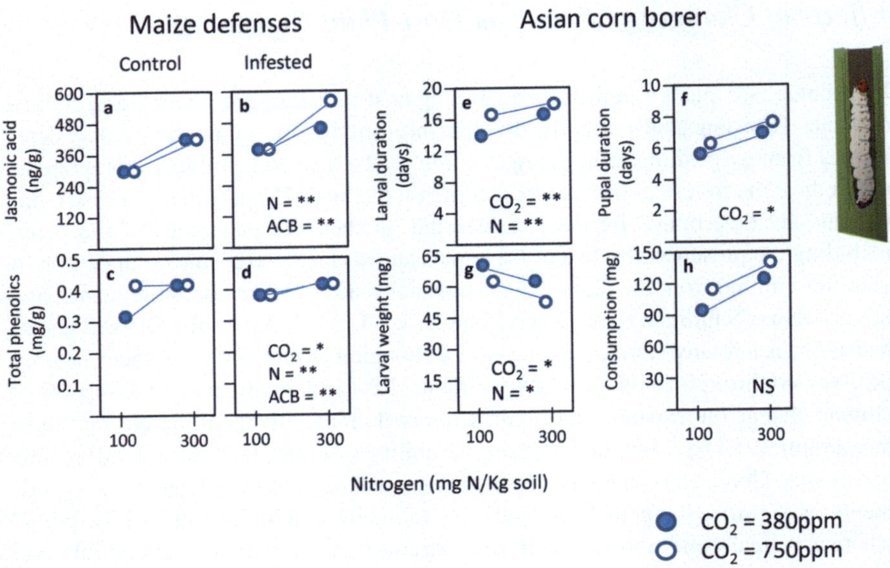

Fig. 10 Separate and combined effects of elevated CO_2 concentration and nitrogen fertilizer on maize defenses (**a–d**) and Asian corn borer (*Ostrinia furnacalis*) development (**e–h**). Jasmonic acid concentrations in (**a**) control uninfested plants and (**b**) plants infested by corn borer and total phenolics in (**c**) control and (**d**) infested plants are indicated; Asian corn borer development is represented by (**e**) larval duration, (**f**) pupal duration, (**g**) larval weight, and (**h**) feeding rate (weight consumed). Results of full factorial analyses are presented as $* = P \leq 0.05$ and $** = P \leq 0.005$ for nitrogen (N), CO_2, and the presence/absence of Asian corn borer (ACB). (Based on data in Xu et al. 2019)

efficiency of food to body weight. Similar results were obtained in experiments with the Asian corn borer, *Ostrinia furnacalis*, feeding on maize. Maize grown under elevated CO_2 had a higher content of defensive phenolics and an increased activity of defensive enzymes (Xu et al. 2019) (Fig. 10). Exposure to elevated CO_2 also reduced fitness during several generations of bollworms, *Helicoverpa armigera*, in experiments with wheat and maize. In each case, the bollworms consumed more foliage to complete their development under elevated CO_2 conditions (Wu et al. 2006; Yin et al. 2010). Rice planthoppers lay more eggs under elevated CO_2 concentrations (Prasannakumar et al. 2012; Shi et al. 2014) (e.g., Fig. 11). Combining elevated CO_2 with higher temperatures also increased planthopper fecundity (by as much as 30%) and increased the rate of nymph development. These effects resulted in 30% more damage to rice under the combined effects of elevated CO_2 and temperature (Pandi et al. 2018). Newman (2004) suggests that elevated CO_2 could increase aphid populations under specific circumstances, including high soil nitrogen (if nitrogen is a limiting factor for aphid development). Surprisingly, the effects of elevated CO_2 on resistant varieties of grain crops (i.e., varieties with known anti-herbivore resistance genes) have not been examined in detail; however, a study by Qian et al. (2019) demonstrated that the resistance of beans against thrips,

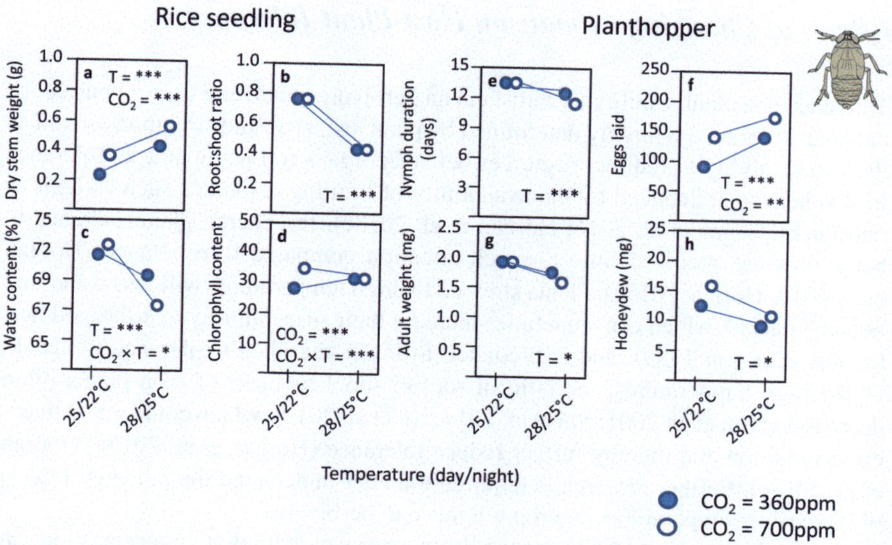

Fig. 11 Separate and combined effects of elevated temperature and CO_2 concentration on rice (**a–d**) and planthopper (*Nilaparvata lugens*) development (**e–h**) in a climate chamber experiment. Seedling development is represented by (**a**) dry shoot weight, (**b**) root/shoot ratio, (**c**) water content, and (**d**) relative chlorophyll content; planthopper development is represented by (**e**) nymph duration, (**f**) eggs laid, (**g**) adult weight, and (**h**) feeding rate (honeydew excreted). Results of full factorial analyses are presented as $* = P \leq 0.05$, $** = P \leq 0.005$, and $*** = P \leq 0.001$ for temperature (T), CO_2, and/or their interactions. (Based on data in Shi et al. 2014)

Frankliniella occidentalis, increased at high CO_2 concentrations due to improved callose deposition.

Other factors related to climate change that influence resistance include light intensity, salinity, and dusts. High light intensity increases the resistance of wheat and maize against the wheat stem sawfly, *Cephus cinctus*, and European corn borer, respectively. This is mainly associated with an increase in the toughness of the stems and leaves of plants grown under high light intensity (Roberts and Tyrrell 1961; Bergvinson et al. 1994, 2012). Salt spray has been shown to reduce plant resistance to aphids, possibly due to increasing plant protein and sugar contents (Braun and Flückiger 1984). Finally dusts can have marked effects on plant-herbivore interactions. Dusts will alter the phylloclimate and may affect photosynthesis, plant respiration, and transpiration (Farmer 1993). Dusts can reduce the effectiveness of defensive plant trichomes (Pelletier et al. 2013; Vanderstock et al. 2019) and deter insect feeding or reduce herbivore fitness through mandible wear or by affecting food digestion (Lucas et al. 2014; LoPresti and Karban 2016). The effects of dusts on cereal resistance to arthropod herbivores deserve increased attention given the increasing networks of non-covered roads through crop-producing regions, particularly in developing countries.

Effects of Changing Climate on Host-Plant Tolerance

Tolerance is a plant's ability to withstand an herbivore attack and to compensate for damage. Tolerance is largely determined by plant age, size, and structure, as well as the plant's ability to redirect resources between organs following attack. Tolerance is also heavily influenced by the availability of limiting resources such as light or nutrients (Horgan et al. 2016; Horgan et al. 2018c). In general, plant tolerance to herbivores has received little research attention compared to resistance (Horgan et al. 2016; Horgan 2017a). Plants stressed by high temperatures will lose water and become flaccid, which can sometimes increase their susceptibility to arthropod herbivores (Larsson 1989), and is predicted to lower the plant's tolerance to attack. Under high temperatures, the ratio of root to shoot biomass of crop plants often decreases (Kim et al. 2001; Shi et al. 2014; Xu et al. 2019), which could affect nutrient acquisition and thereby further reduce tolerance (Horgan et al. 2016a; Horgan et al. 2018c). Further research is required to better understand the potential effects of increasing temperatures on crop tolerance to herbivores.

In C3 plants, elevated CO_2 generally increases carbohydrate concentrations in the plant and therefore increases the carbon/nitrogen [C/N] ratio (demanding increased food consumption by herbivores) (Kim et al. 2001; DeLucia et al. 2012; Xu et al. 2019). However, the effects of elevated CO_2 on herbivores are strongly dependent on the availability of nitrogen (Newman 2004). Despite increased food consumption rates, herbivores on crop plants are unlikely to be limited by low nitrogen availability (i.e., high C/N ratios) because crops normally receive extra nitrogen in the form of organic or synthetic fertilizers. Therefore, the predicted negative effects of elevated CO_2 on plant tolerance will probably be lower for crop plants than for non-crop species. The effects of nitrogenous fertilizers will also vary between herbivore species. In general, plant-sucking insects can more effectively acquire nitrogen from the plant phloem and will be less affected by elevated CO_2 than chewing insects (Horgan et al. 2016a). Because nitrogen increases plant tolerance to herbivores (Horgan et al. 2016; Horgan et al. 2018c), then high C/N ratios in plant tissues and competition for nitrogen between the plant and its herbivores might be expected to reduce plant tolerance in some cases. Higher damage to leaves per unit weight of herbivore and the availability of CO_2 in the atmosphere might also improve plant tolerance in some cases (particularly for chewing insects). For example, in a non-crop system, herbivore damage to leaves of silver birch increases the uptake of CO_2, particularly under high nutrition and at higher temperatures, which has a positive impact on the ability of birch seedlings to compensate for herbivore damage (Huttunen et al. 2007). In an experiment with maize, a C4 plant, elevated CO_2 alone (i.e., without high temperatures) increased plant starch content, but did not affect the C/N ratio [although this differed from previous results by the same authors: Xu et al. (2019)]. Without greatly affecting C/N ratios, elevated CO_2 is therefore unlikely to affect the tolerance of maize and other C4 crops to herbivore damage.

Effects of Changing Climate on Host-Plant Vulnerability

Vulnerability is largely determined by the plant's exposure to herbivore pressure. Synchrony in the timing of attack by pests with favored stages of crop development is a key aspect in the vulnerability of crop plants to herbivores. Warmer global temperatures might increase or decrease synchrony, depending on the crop and herbivore species in question (Forrest 2016; Wu et al. 2019). Unlike natural vegetation, the phenology of crop plants is largely determined by sowing dates as decided by farmers. This may be influenced by the capacity of different varieties to tolerate extremes in temperature or light intensity. At higher latitudes, early sowing can reduce herbivore damage by permitting crops to escape peaks in herbivore migration or by avoiding maximum egg hatch or the break of cold-induced herbivore diapause. The importance of escaping herbivory has been recognized for decades (Jacobson and Farstad 1952; Buntin et al. 1990). For example, damage to wheat from the wheat bulb fly, *Delia coarctata*, is strongly influenced by sowing date. Early sowing reduces vulnerability of the crop to the fly by advancing crop development beyond the most susceptible stages before flies attack (Jones 1978). Late sowing can also reduce the survival of pest populations. For example, late-sown maize is associated with low levels of diapause in *Sesamia nonagrioides* in Spain leading to higher mortality of the pest during winter months (Eizaguirre et al. 2007). Careful monitoring of pest populations can therefore help farmers to better manage the timing of crops and disrupt synchrony between pests and favored growth stages under changing climatic conditions [e.g., rice planthoppers in Asia (Horgan 2017a)]. However, nonmigratory pest species can adapt to changing crop phenologies to maintain synchrony with their main crop hosts or with regional peaks in planting (Krysan et al. 1984; Levine et al. 1992a).

Warmer climates could also increase crop vulnerability by altering the dynamics of herbivore-vectored plant diseases. For example, high temperatures increase the acquisition of rice waika virus by *Nephotettix cincticeps*, thereby potentially increasing viral transmission (Inoue and Hirao 1980). In contrast, high rates of transovarial passage of rice stripe virus (>90%) in the small brown planthopper, *Laodelphax striatellus*, are associated with cooler temperatures such that viral transmission is reduced under warmer climates (Raga et al. 1988). Avoiding long-duration varieties will reduce crop vulnerability to herbivores and their viruses but could be countered by increasing numbers of generations in multivoltine herbivore species.

Indirect Biodiversity-Mediated Climate Effects on Herbivores

Crop herbivores are part of complex agroecosystems that include a wide variety of natural enemies including entomopathogenic bacteria, fungi and nematodes, parasitic wasps and flies, and predatory spiders, bugs, and beetles (Settle et al. 1996;

Dainese et al. 2019). The arthropod natural enemies of herbivore pests may respond directly to climate change by expanding their distribution ranges, by increasing the number of generations they complete in a season, and by changes in their overwintering ranges, migration times, and migration paths in the same way as occurs with herbivore species (Sect. 5). A key question is whether such responses by natural enemies will be sufficient to counter any potential positive effects of climate change on insect herbivore populations and thereby reduce the risks of increased crop damage. In recent years, there has been an increase in the number of climate-related studies that include three or more trophic levels (Hoover and Newman 2004; Schuldiner-Harpaz and Coll 2013; Abbott et al. 2014; Barton 2014; Barton and Ives 2014) – particularly in crop-aphid-parasitoid systems (Bensadia et al. 2006; Guay et al. 2009; Dong et al. 2013). Studies such as these can help predict the efficiency of natural enemies in responding to climate-induced changes in herbivore populations and assemblages.

Range Expansion and Voltinism in Arthropod Natural Enemies

The maintenance of crop herbivores at low population densities often requires specialist natural enemies that track changes in herbivore populations and display density-dependent attack responses (Horgan 2017b; Dainese et al. 2019; Damien and Tougeron 2019). However, generalist natural enemies also play a key role in herbivore population dynamics (Settle et al. 1996; Horgan et al. 1999; Horgan and Myers 2004). For example, in rice, generalist natural enemies such as spiders and predatory bugs protect early crop stages from herbivore attack by initially feeding on decomposers and thereby maintaining high numbers in the rice paddies prior to rice planting (Settle et al. 1996). Natural biological control therefore relies on both specialists and generalists. For example, invasive species during post introduction outbreaks will often experience high mortality from generalist predators, but are not regulated by any specialist natural enemies and therefore display unstable population dynamics (Horgan et al. 1999; Horgan and Myers 2004; Moreau 2006). This suggests that crop herbivores that expand their distribution ranges in response to climate warming could display similar instability in their populations unless specialist natural enemies are already present in the new environment. Spiders tend to be generalist predators, and although they can disperse long distances using thermal drafts and jet streams (Riley et al. 1994, 1995; Sedlock et al. 2019), they probably have normally lower rates of range expansion than insect herbivores. Furthermore, spiders and other generalists already present in the expanded range will probably cause similar rates of herbivore mortality as spiders in the original range. Such generalists, although they contribute significantly to herbivore mortality, are often ineffective in regulating herbivore populations.

Specialists are highly mobile species and generally have high reproductive rates including polyembryony in some species (Hance et al. 2006; Jeffs and Lewis 2013). Specialist parasitoids may have difficulty in tracking their hosts over expanding ranges, particularly in the short term when host populations are likely to be patchily distributed (Horgan et al. 1999; Jeffs and Lewis 2013). However, there are also possibilities that specialist natural enemies spreading poleward or to higher elevations could more heavily impact already widespread herbivores. For example, the parasitoid *Sturmia bella* that was introduced into the UK has been associated with a decline in populations of the non-pest butterfly *Aglais urticae* over the parasitoid's introduced range (Jeffs and Lewis 2013). Increasing voltinism in herbivores under warmer climates could have negative or positive effects on specialist parasitoids. For example, extra generations of host herbivores could offer a greater resource and extended time for parasitoids to build up populations (Jeffs and Lewis 2013). Alternatively, parasitoids might be prone to ecological traps where their herbivore hosts are poorly synchronized with the crop plant or where parasitoids become trapped during diapause in their insect host. Some parasitoids, particularly koinobiont parasitoids (that allow the host to continue feeding after infested), display a high degree of plasticity in their development rates and final body size (Klapwijk et al. 2010), which should allow them to rapidly adapt to climate-induced changes in the population dynamics and/or quality of their herbivore hosts. However, there is also evidence that compared to their hosts, some specialist parasitoids may be less plastic in their ability to acclimatize to temperature extremes. For example, the stem borer *Chilo partellus* displays greater plasticity compared to its parasitoid *Cotesia flavipes* during temperature acclimation studies (Mutamiswa et al. 2018).

Under the future high temperatures predicted by climate models, generalist predators and established parasitoids will also increase the number of generations they complete. Research from Japan suggests that many natural enemy species will increase their generations by two or three generations under a mean monthly temperature increase of 2 °C, whereas most rice pests will only increase their generations by one extra generation (Kiritani 1999). This suggests that unlike other invasive herbivores, in response to climate change, widespread crop pests are unlikely to display unstable population dynamics in expanded ranges and under increased voltinism if natural enemies are already present and are adequately conserved.

In Situ Responses by Natural Enemies to a Changing Climate

The natural enemies of crop pests display a wide range of physiological and behavioral responses to climate extremes. These can result in an increased or decreased efficiency in predation or parasitism and potentially affect regulation and the stability of herbivore populations (Chiverton 1988; Runjie et al. 1996; Clercq 2001; Jamshidnia et al. 2010; Ma et al. 2012). Synchrony between natural enemies and their herbivore prey/hosts is a key determinant of regulatory efficiency. Insect herbivores may exhibit poor synchrony with favored stages of their host plant under an increasingly variable climate (Sect. 6.3); however, for parasitoids that depend on environmental, herbivore, and herbivore damage-related cues to guide activity cycles and find their hosts, poor synchrony is likely to be even more prominent. For example, in an experiment from China, although population growth of the aphid *Sitobion avenae* was accelerated in experimentally warmed wheat plots, parasitoids failed to respond to high aphid densities in one of 2 years (Dong et al. 2013). These results may be due to the small scale of the field-plot experiments; indeed, the aphid and its parasitoids have been shown to be normally highly synchronized over larger scales (Jeffs and Lewis 2013; Pérez-Rodríguez et al. 2015). Nevertheless, the study does indicate the potential effects of herbivore-parasitoid mismatches in driving pest population growth under a future, more variable climate.

Several studies suggest that natural enemies may display increased efficiency under elevated temperatures. For example, ladybeetles (Coleoptera: Coccinellidae) and carabids (Coleoptera: Carabidae) were more effective in reducing aphid numbers at high temperatures because of shorter prey handling times (Abbott et al. 2014) and increased searching efficiency (Chiverton 1988), respectively. The parasitoid *Cotesia flavipes* had higher rates of parasitism on the pink stem borer, *Sesamia inferens*, at higher temperatures in studies from rice fields in Japan (Mia and Iwahashi 1999), and the predatory bugs *Podisus maculiventris* and *Podium nigrispinus* had improved attack efficiency on the beet armyworm, *Spodoptera exigua*, under higher temperatures (Clercq 2001). In rice, the parasitoid *Cardiochiles philippinensis* had increased functional efficiency when feeding on leaf folders (Runjie et al. 1996), and the mirid predator *Cyrtorhinus lividipennis* and hymenopteran parasitoid *Anagrus nilaparvatae* both had greater efficiency when feeding on planthopper eggs under high temperatures (Song and Heong 1997; Ma et al. 2012). Herbivores may also become more vulnerable to predators where they avoid high temperatures (Roitberg and Myers 2012). For example, under hot weather conditions, aphids drop to the soil where they become more vulnerable to edaphic predators. Studies with entomopathogens, including entomopathogenic nematodes, have also indicated higher rates of herbivore mortality at elevated temperatures (Stacey and Fellowes 2007; Gözel and Güneş 2013).

It is still unknown how increased natural enemy efficiency, mainly reported from laboratory studies, might affect herbivore populations under warmer or more variable climates. Unsurprisingly, some of the same studies mentioned above have also indicated that the efficiency of parasitoids and other natural enemies is reduced under temperature extremes. For example, mortality of the rice leaf folder due to *Cardiochiles philippinensis* and mortality of the beet armyworm, *Spodoptera exigua*, from the predators *Podisus maculiventris* and *Podisus nigrispinus*, declined at temperatures above 28 °C (Runjie et al. 1996; Clercq 2001). Several studies have also indicated that hymenopteran parasitoids in particular are highly negatively affected by temperature extremes (Thomson et al. 2010). The ultimate effects of temperature-related increases in natural enemy efficiency probably depend more on aspects of the herbivore and natural enemy life cycles than mortality rates. For examples, at temperatures below 11 °C, the reproductive rate of the aphid *Acyrthosiphon pisum* falls below the predation rate of the ladybeetle *Coccinella septempunctata* (Hullé et al. 2010), such that the ladybeetle is more effective in reducing aphid populations at lower temperatures. Furthermore, lower soil moisture contents associated with higher temperatures could reduce the efficacy of entomopathogenic nematodes, fungi, or bacteria. A meta-analysis of parasitism on (non-pest) lepidopterans has indicated a strong relation between variability in precipitation and parasitism rates, mainly due to a reduction in parasitism by hymenopterans (Stireman et al. 2005).

To gain increased inference, future studies of natural enemy efficiency could be conducted over longer time scales and more natural conditions, and they could include multiple predator and prey generations. How differential responses to climate by herbivores and their natural enemies might affect herbivore population regulation will be of particular interest. To date, experimental studies on the potential effects of climate change on herbivore population dynamics are rare. However, Meisner et al. (2014) recently indicated that high temperatures that increased the fitness of an aphid and its parasitoid *Aphidius ervi* and increased the attack efficiency of the parasitoid were translated into complex effects on population cycling in a multigeneration experiment. The authors indicated that increasing temperatures would actually destabilize aphid populations by increasing the amplitude but decreasing the period of population cycles.

Responses by herbivores and their natural enemies to climate change are often complex, and the ultimate results of such changes can sometimes appear counterintuitive. For example, in a field experiment with corn leaf aphids, *Rhopalosiphum maidis*, the aphid-tending ant *Prenolepis imparis*, and the predatory ladybeetle *Harmonia axyridis*, aphid abundance increased on corn under elevated temperatures but declined in the presence of the ladybeetle (Fig. 12a). Aphid-tending ants protect aphids from predators, receiving a food reward (honeydew excreted by the aphids) in return (Fig. 12 b,c). Under ambient temperatures, the ants could sufficiently fend off predators to maintain aphid populations; however, at experimentally elevated temperatures, ants were incapable of protecting the aphids because of a reduction in defense efficiency (Fig. 12c). In effect, warming had broken down the ant-aphid mutualism leading to declines in aphid population growth under higher temperatures (Barton and Ives 2014). In a similar multispecies interaction, clones of the pea aphid, *Acyrthosiphon pisum*, that contain the endosymbiont *Hamiltonella defensa*, are immune to attack by the parasitoid *Aphidius ervi* at 20 °C (but clones without the endosymbiont are not). The same symbiont-bearing clones were highly susceptible to the parasitoid at higher temperatures, presumably due to heat sensitivity of the endosymbiont (Bensadia et al. 2006; Guay et al. 2009; Jeffs and Lewis 2013).

A few studies have examined the effects of other global changes on the natural enemies of crop pests. Elevated CO_2 had negligible effects on predation by the ladybeetle *Harmonia axyridis* on the aphid *Sitobion avenae* feeding on wheat, but the abundance of a braconid parasitoid, *Aphidius picipes*, was higher under elevated CO_2, resulting in increased aphid parasitism under elevated compared to ambient CO_2 (Chen et al. 2007; Thomson et al. 2010). Elevated CO_2 could affect the development of koinobiont parasitoids by reducing the quality of the insect host; however, the extended development time of herbivores feeding on plants under high CO_2 and greater damage by herbivores to host foliage could also increase parasitism rates by extending the time available to attack the herbivore and by enhancing host-

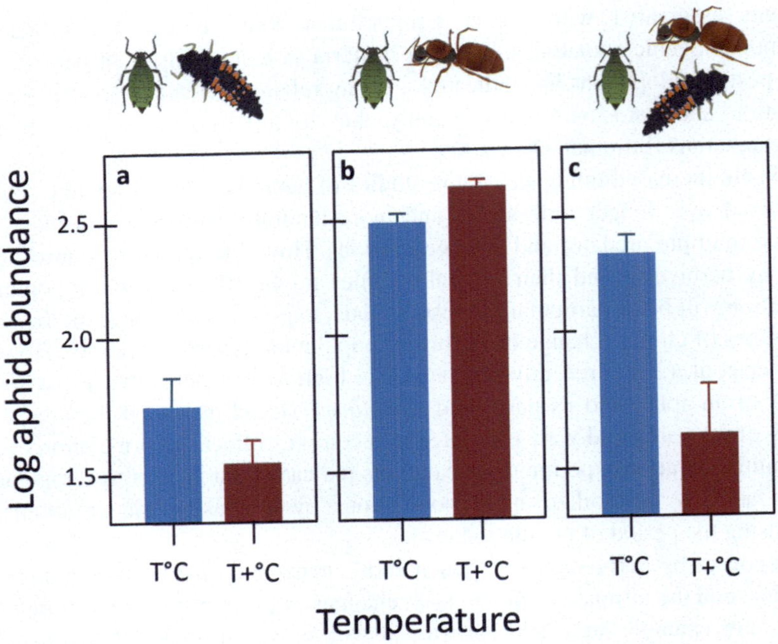

Fig. 12 Abundance of corn leaf aphids, *Rhopalosiphum maidis*, in field plots with (**a**) the predatory ladybeetle *Harmonia axyridis* and (**b**) the aphid-tending ant *Prenolepis imparis* at ambient (T°C: blue bars) and elevated (T + °C: red bars) temperatures. The relative abundance of the aphids in the presence of both the ants and ladybeetles is presented in (**c**). Note that aphid-tending ants protected aphids from predation at ambient temperatures but were incapable of protecting aphids at elevated temperatures. (Figure drawn using data presented by Barton and Ives 2014)

finding cues. Finally, in a field study by Barton (2014), predation of aphids by the ladybeetle *Harmonia axyridis* was higher under reduced wind conditions, as predicted for future climates, because of a reduction in the movement of plants. Blocking wind in experimental field plots resulted in a 40% decline in aphid abundance compared to control plots.

Indirect Crop Management-Mediated Climate Effects on Arthropod Herbivores

In Sect. 6.3, crop management and host-plant ontogeny (age-related changes in plant physiology) were seen to interact and potentially determine crop vulnerability to insect herbivores under a changing climate. Much of these effects could be mitigated by altering sowing or planting times, adjusting crop duration, or scheduling crop management activities based on the results of local or regional herbivore monitoring programs. Crop nutrient management (i.e., the application of nitrogenous fertilizers) was also seen to have a major influence on the resistance and tolerance of crops to arthropod herbivores and could counter the potential reducing effects of elevated CO_2 on herbivore populations. In this section, the potential impacts of other crop management activities will be addressed. The effects of these activities are not plant-mediated, or where they are plant-mediated, the activities are practiced by farmers primarily as a method to control pest damage. These activities include landscape modification, the application of fertilizers and other soil amendments, and the application of pesticides.

Climate change has already affected the way that farmers manage their crops. Farmers in many regions have adopted crops that are better suited to their recent, local climatic conditions (climate adaptation) (Reidsma et al. 2010; He et al. 2015). This includes shifts toward increasing production of forage maize and warm winter crops in northern Europe (Olesen et al. 2011), changes in crop rotation and other cropping practices (Nelson et al. 2009; Kuntashula et al. 2014), moves to drought- or flood-tolerant crop varieties (Yamano et al. 2013; Singh 2017), or the abandonment of land due to salinity intrusion or soil deterioration (Singh and Singh 1995; Horgan et al. 2018b). Increasing variability in weather may also cause some farmers to abandon crops as conditions change within a growing season (Fig. 13). Changes in climate occur at the same time as other global changes (e.g., increases in mechanization, increased urbanization, and increases in the production and use of agrochemicals), and many of these will also affect crop management. To assess the impact of climate change on crop production and herbivore pressures, some statistical models have included factors related to crop management such as planting times, cropping patterns, landscape heterogeneity and composition, or fallow management (Hu et al. 2011; Hu et al. 2014; Zhao et al. 2016). Models have been less likely to include predominant crop varieties, the number of pesticide applications and the types of pesticides used, or predominant fertilizer regimes. This data is often

Fig. 13 Climatic variability. Farmers in Java, Indonesia, plant rice and maize in a paddy field as they struggle to anticipate the best crop for the coming season's weather conditions (Photo F. Horgan)

difficult to come by, but such factors could be key to causing major shifts in herbivore dynamics.

Changes in landscape structure can have significant effects on arthropod communities (Zhao et al. 2016; Sánchez-Bayo and Wyckhuys 2019). Changes in landscape have been associated with often phenomenal declines in arthropod biodiversity (Sánchez-Bayo and Wyckhuys 2019). Because many herbivore pests are monophagous or oligophagous species that specialize on crops, landscape is unlikely to greatly influence their life cycles and development. Changes in landscape will, however, affect many species that are the natural enemies of important crop pests (Gurr et al. 2016; Horgan et al. 2017b, 2017c; Dominik et al. 2017, 2018; Dainese et al. 2019). Whereas many of these are also more common in open field crops than in native vegetation, others rely on alternative food resources to sustain their populations during times when crop pests are not active (i.e., before migrants arrive) or where crops are not yet planted (Settle et al. 1996; Gurr et al. 2016). For example, mortality of grain crop pests due to hymenopteran parasitoids has been associated with the availability of nectar-producing flowers in the landscape that increase parasitoid fecundity and attack efficiency (Gurr et al. 2016; Vu et al. 2018; Horgan et al. 2019a). Pest mortality due to natural enemies often declines at greater distances from non-crop vegetation (Gurr et al. 2016; Vu et al. 2018). A study by Zhao et al. (2016) highlighted that damage to crops from six different herbivore pests in China between 1951 and 2010 increased significantly with increasing areas of arable land in the landscape. Only damage from two pest species, the rice leaf folder and the armyworm, *Mythimna separata*, was more closely associated with climate; however, in both cases, these species caused less damage under higher mean temperatures (Zhao et al. 2016). Linked with the expansion of arable land and the consequent reduction in spatial landscape complexity, farmers in many regions have also increased the number of crops they produce each year – thereby also reducing temporal complexity. For example, improved irrigation in many parts of Southeast Asia has allowed farmers to plant an extra rice crop each year (i.e., double or triple cropping) (Qiu et al. 2003; Sakamoto et al. 2006). In some parts of Indonesia, farmers can plant seven rice crops every 2 years (personal observation). Crop-free fallow periods are important to break pest and disease cycles (Horgan 2017a). By eliminating fallows and expanding crop areas, farmers will exacerbate the effects of increased voltinism in pest species under warmer climates.

It is reasonable to predict that because of the widespread distribution of most crop pests, the addition of nutrients to crop soils, the role of generalist predators in suppressing pest populations, and the parallel responses by pests and natural enemies to changing climate conditions, crop damage from most pests will not appreciably increase under future climates (Fig. 14). However, where the diversity and abundance of natural enemies are reduced, pest populations could exhibit large increases in outbreak severity and frequency. There is now considerable evidence to indicate that pest outbreaks in grain crops are frequently the result of poor crop management including excessive fertilizer use, a poor choice of crop varieties, and the use, misuse, or overuse of pesticides, particularly insecticides (Godfray and Chan 1990; Hardin et al. 1995; Guedes and Cutler 2014; Horgan et al. 2017c).

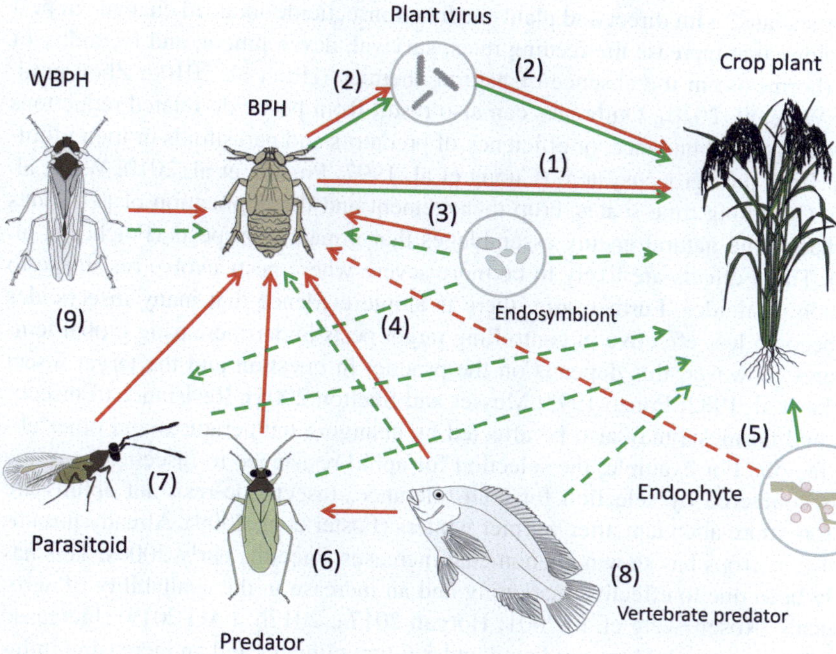

Fig. 14 Potential effects of climate change on herbivory in a complex agroecosystem. [1] Evidence suggests that the brown planthopper (*Nilaparvata lugens*: BPH) could complete a greater number of generations and cause increasing damage (indicated by a red arrow) to rice under a warmer (+2 °C) climate and elevated CO_2 (>500 ppm) (Shi et al. 2014) – although this depends on the geographical location of the population – and could result in declining populations at lower latitudes (indicated by a green arrow) (Sujithra and Chander 2013). [2] Planthoppers may also acquire and transmit plant viruses at a greater (green arrow) or lesser (red arrow) rate (Inoue and Hirao 1980; Raga et al. 1988). [3] Yeast-like endosymbionts in the planthopper abdomen are sensitive to high temperatures and could delay planthopper growth and development (red arrow) which indirectly benefits the rice plant (broken green arrow) (Ferrater et al. 2015). [4] Endosymbionts may also determine levels of parasitism – as occur in aphids (Bensadia et al. 2006) – with high temperatures benefiting parasitoids. [5] Endophytes can have increased activity under higher temperatures directly promoting crop development (green arrow), increasing resistance, and indirectly (broken red arrow) reducing planthopper fitness as detected in other plant-herbivore systems (Salminen et al. 2005). [6] Arthropod predators like *Cyrtorhinus lividipennis*, [7] parasitoids like *Anagrus* spp., and [8] vertebrate predators will have greater efficiency in attacking planthopper, and many will complete a greater number of generations than the planthoppers during the growing season (Song and Heong 1997; Kiritani 1999; Ma et al. 2012), thereby indirectly protecting the crop. [9] Temperature-dependent and CO_2-dependent competition can alter the outcome of interspecific competition [as occurs with some lepidopteran pests (Ntiri et al. 2016)] potentially benefiting brown planthoppers at low competitor densities [through feeding induction (Srinivasan et al. 2016)] but reducing populations in favor of a less virulent species at high competitor densities

Among the best-studied cases of management-induced outbreaks are those relating to rice pests – including planthoppers, leaf folders, and stem borers – that reach high densities in the enemy-free space created by broad-spectrum insecticides (Godfray and Chan 1990; Horgan et al. 2017c; Wu et al. 2020). Outbreaks of these pests have been associated with direct and plant-mediated insecticide-induced changes in pest physiology that increase the feeding rates, survival, development, and fecundity of pests (hormesis) in the absence of natural enemies (Hu et al. 2010c; Zhou et al. 2019; Wu et al. 2020). Outbreaks can also result from pesticide-related reductions in the diversity, abundance, or efficiency of predators and parasitoids or their elimination from the agroecosystem (Cuong et al. 1997; Preetha et al. 2010; Wu et al. 2020). Over long time scales, crop management and the application of pesticides can change the natural enemy assemblages that impact crop pests (Folcher et al. 2011). These effects are likely to be more severe where pests evolve resistance to common pesticides. Furthermore, there is ample evidence that many insecticides may become less effective in controlling target pests under increasing global temperatures; however, this depends on the product in question and the target insect (Sparks et al. 1982; Punzo 1993; Musser and Shelton 2005). Resistance to insecticides and hormesis may also be affected by changing temperatures and other climate factors. For example, the selection for aphid resistance to insecticides in the UK is countered by selection for cold tolerance. Insecticide-resistant aphids are therefore more abundant after warmer winters (Foster et al. 2009). Already insecticide use in crops has shown phenomenal increases since the early 2000s. This has largely been due to effective marketing and an increase in the availability of agrochemicals (Rosenzweig et al. 2001; Horgan 2017a, 2017b; FAO 2019). Increased variability in crop yields as predicted under future climates and an increasing influence of the private sector in discourse around climate change (Capstick et al. 2015; Farrell 2016) could sway farmers' behaviors toward even further increases in pesticide use. These effects will greatly destabilize crop pest populations under global climate change.

Crop Management for Future Climates

Although the potential effects of a changing global climate on pest population pressure is often predicted to increase under current models, the ultimate effects of climate change on actual damage to crops has been less predictable because of indirect climate effects (Fig. 14). To counter potential increases in herbivore population pressure, future crop production systems should focus on holistic approaches to crop management and avoid simplistic pest management practices – such as advocating or increasing insecticide use in response to models of temperature-related increases in pest range or voltinism. Indeed, as outlined in Sect. 8, inappropriate pest management solutions that destabilize pest populations could inadvertently lead to increased insect outbreaks under warmer climates. Crop managers will therefore need to incorporate practices that stabilize herbivore populations, such as reducing insecticide use, promoting conservation biological control, and optimizing indirect plant-mediated climate effects by adopting climate-resilient plant resistance, promoting plant tolerance, and reducing crop vulnerability.

Host-plant resistance can be promoted by eliminating highly susceptible crop varieties from the environment. For example, currently a large proportion of hybrid rice varieties is highly susceptible to the white-backed planthopper (Horgan et al. 2016) and should be replaced by increasingly resistant hybrids. Resistant varieties should be developed with anti-herbivore resistance genes that are not sensitive to temperature or CO_2 concentrations (Wang et al. 2010). Resistance can be further increased in grain crops by using a range of soil amendments including potassium, silicon, and biochar. Silicon and biochar promote a general resistance in crops against a range of diseases and arthropod pests (Kvedaras et al. 2010; Shakoor 2014; Bakhat et al. 2018; Nascimento et al. 2018). Furthermore, silicon levels in the soil can be augmented by applying crushed, fast-reacting silicate rocks that could also moderate climate change by actively removing CO_2 from the atmosphere (Shakoor 2014; Beerling et al. 2018). Similarly, the application of biochar has multiple benefits for crop production while also improving crop resistance to pests (Tang et al. 2013; Chen et al. 2019). Biochar improves soil structure and promotes soil health. Biochar also sequesters carbon, thereby reducing CO_2 emissions (Tang et al. 2013; Beerling et al. 2018). Crop tolerance to arthropod pests can be increased by ensuring that plants have optimal conditions for modular growth. This is accomplished by ensuring ample resource availability, reducing intraspecific competition (e.g., by increasing plant spacing), and achieving optimal plant architecture. For example, many of the practices promoted through the system of rice intensification (SRI) are directed toward achieving plants that are tolerant to biotic and abiotic stresses (Uphoff 1999). A focus on large, healthy root systems in SRI will counter CO_2-related and temperature-related reductions in root-shoot ratios (Prasannakumar et al. 2012; Shi et al. 2014; Xu et al. 2019) and thereby increase plant nutrient acquisition and promote tolerance to herbivory. Finally, crop vulnerability can be reduced by avoiding long-duration varieties and avoiding synchrony between planting activities and peaks in the abundances of key pests (Horgan 2017a).

The regulatory efficiency of natural enemies must be properly managed to optimize indirect biodiversity-mediated climate effects on crop health. Healthy soils that promote endophytes can improve host-plant resistance and directly deter or kill pest herbivores. Healthy soils will also support populations of entomopathogenic nematodes and other microorganisms. Soil health can be improved by using biochar, reducing tillage practices, or providing ground cover in the form of mulch or green manures (Hooks and Johnson 2004; Schmidt et al. 2004). Ground cover also prevents evaporation and conserves soil moisture (Manns et al. 2007). At field scales, reducing pesticide use or eliminating pesticides altogether will promote natural pest regulation as well as provide several further advantages for human and environmental health. Insecticides can often be avoided without consequence, or they can be replaced using alternative pest management practices such as biological control (Horgan 2017a, 2017b; Babendreier et al. 2019). Replacing chemical insecticides with organic insecticides will not reduce the potential for instability in the agroecosystem because organic pesticides can also kill beneficial arthropods, including natural enemies (Horgan et al. 2018a). Biological control should be climate-resilient. A number of studies indicate that cold storage of biological control agents can affect their efficiency at higher temperatures; therefore, some screening for climate resilience and optimization of rearing conditions to best suit local and changing climates is warranted (Yuan et al. 2012; Babendreier et al. 2019).

Among the most promising approaches for attaining crop resilience against climate change are a range of agroecological field-based and landscape-based practices aimed at providing functional habitats for natural enemies (Gurr et al. 2016; Horgan et al. 2017b, 2017c; Heeb et al. 2019). These practices can be regarded as part of a biodiversity management that reduces habitat suitability for pests but promotes natural enemies and other beneficial ecosystem services. Much of the approach is based on planting functional plants in and around the agroecosystem. Plants are mainly selected based on their capacity to deter or kill pests and attract or sustain natural enemies (Gurr et al. 2016; Heeb et al. 2019). However, functional plants may also provide alternative sources of income for farmers or increased nutrient diversity for farming households (Horgan et al. 2017a, 2017b; Horgan et al. 2019a). Furthermore, functional plants may remediate soil contamination by heavy metals and other environmental pollutants (Parker and Page 1994; Tang et al. 2013) while at the same time providing resources for the natural enemies of crop pests. At larger scales, attention to the connectivity of native vegetation or patches of functional plants (either as vegetation strips or plots) can improve natural enemy efficiency in finding and handling their pest prey/hosts or in surviving during periods of low pest abundance (Dominik et al. 2017, 2018). Importantly, biodiversity management will increase the biodiversity of agroecosystems, thereby conserving threatened vertebrate and invertebrate species in agroecosystems and providing habitat for dispersing populations (Horgan et al. 2017b; Sedlock et al. 2019). Whereas pesticides have been shown to reduce the complexity of arthropod communities and promote instability (Cohen et al. 1994; Schoenly et al. 1996), agroecology and crop diversification will increase food-web complexity to provide stability and resilience (Ings et al. 2009). In the face of insufficient knowledge to predict the ultimate outcomes of the sum of indirect biodiversity-mediated climate effects on herbivores, the promotion of ecosystem resilience and arthropod community stability through agroecology and sustainable farming practices should be regarded as a priority for future crop production systems.

References

Abbott KC, Harmon JP, Fabina NS (2014) The challenge of predicting temperature effects on short-term predator–prey dynamics. Popul Ecol 56:375–392

Ali M, Huang D, Nachman G, Ahmed N, Begum MA, Rabbi M (2014) Will climate change affect outbreak patterns of planthoppers in Bangladesh? PLoS One 9:e91678

Ali MP, Kabir MMM, Afrin S, Nowrin F, Haque SS, Haque MM et al (2019) Increased temperature induces leaffolder outbreak in rice field. J Appl Entomol 143:867–874

Babendreier D, Wan M, Tang R, Gu R, Tambo J, Liu Z et al (2019) Impact assessment of biological control-based integrated pest management in rice and maize in the Greater Mekong Subregion. Insects 10:e226

Baccini A, Goetz SJ, Walker WS, Laporte NT, Sun M, Sulla-Menashe D et al (2012) Estimated carbon dioxide emissions from tropical deforestation improved by carbon-density maps. Nat Clim Chang 2:182–185

Bakhat HF, Bibi N, Zia Z, Abbas S, Hammad HM, Fahad S (2018) Silicon mitigates biotic stresses in crop plants: a review. Crop Prot 104:21–34

Bale JS, Hayward SAL (2010) Insect overwintering in a changing climate. J Exp Biol 213:980–994

Bale JS, Masters GJ, Hodkinson ID, Awmack C, Bezemer TM, Brown VK et al (2002) Herbivory in global climate change research: direct effects of rising temperature on insect herbivores. Glob Chang Biol 8:1–16

Barton BT (2014) Reduced wind strengthens top-down control of an insect herbivore. Ecology 95:2375–2381

Barton BT, Ives AR (2014) Direct and indirect effects of warming on aphids, their predators, and ant mutualists. Ecology 95:1479–1484

Battisti A, Larsson S (2015) Climate change and insect pest distribution range. In: Björkman C, Niemelä P (eds) Climate change and insect pests. CABI, Wallingford, pp 1–15

Beerling DJ, Leake JR, Long SP, Scholes JD, Ton J, Nelson PN et al (2018) Farming with crops and rocks to address global climate, food and soil security. Nature Plants 4:138–147

Bensadia F, Boudreault S, Guay J-F, Michaud D, Cloutier C (2006) Aphid clonal resistance to a parasitoid fails under heat stress. J Insect Physiol 52:146–157

Bergvinson DJ, Arnason JT, Hamilton RI, Tachibana S, Towers GHN (1994) Putative role of photodimerized phenolic acids in maize resistance to *Ostrinia nubilalis* (Lepidoptera: Pyralidae). Environ Entomol 23:1516–1523

Bergvinson DJ, Larsen JS, Arnason JT (2012) Effect of light on changes in maize resistance against the European corn borer, *Ostrinia nubilalis* (Hübner). Can Entomol 127:111–122

Bosque-Pérez NA (1979) Major insects pests of maize in Africa: biology and control. International Institute of Tropical Agriculture, Ibadan

Brabec M, Honěk A, Pekár S, Martinková Z (2014) Population dynamics of aphids on cereals: digging in the time-series data to reveal population regulation caused by temperature. PLoS One 9:e106228

Braun S, Flückiger W (1984) Increased population of the aphid *Aphis pomi* at a motorway. Part 2 The effect of drought and deicing salt. Environ Pollut Ser A Ecol Biol 36:261–270

Brohan P, Kennedy JJ, Harris I, Tett SF, Jones PD (2006) Uncertainty estimates in regional and global observed temperature changes: a new data set from 1850. J Geophys Res Atmos 111:D12

Buntin DG, Bruckner PL, Johnson JW (1990) Management of Hessian fly (Diptera: Cecidomyiidae) in Georgia by delayed planting of winter wheat. J Econ Entomol 83:1025–1033

Byers DL (2017) Studying plant–pollinator interactions in a changing climate: a review of approaches. Appl Plant Sci 5:e1700012

Cannon RJ (1998) The implications of predicted climate change for insect pests in the UK, with emphasis on non-indigenous species. Glob Chang Biol 4:785–796

Capstick S, Whitmarsh L, Poortinga W, Pidgeon N, Upham P (2015) International trends in public perceptions of climate change over the past quarter century. Wiley Interdiscip Rev Clim Chang 6:35–61

Challinor AJ, Watson J, Lobell DB, Howden SM, Smith DR, Chhetri N (2014) A meta-analysis of crop yield under climate change and adaptation. Nat Clim Chang 4:e287

Chen FJ, Wu G, Parajulee MN, Ge F (2007) Impact of elevated CO_2 on the third trophic level: a predator *Harmonia axyridis* and a parasitoid *Aphidius picipes*. Biocontrol Sci Tech 17:313–324

Chen C-Y, Lai C-Y, Kuo M-H (2009) Temperature effect on the growth of *Buchnera* endosymbiont in *Aphis craccivora* (Hemiptera: Aphididae). Symbiosis 49:53

Chen X, Liu W, Zhu J, Zhang X, Zhai B (2013) Climatic factors influencing the 2003 outbreak of *Cnaphalocrocis medinalis* in China. Chin J Appl Entomol 50:615–621

Chen Y, Li R, Li B, Meng L (2019) Biochar applications decrease reproductive potential of the English grain aphid *Sitobion avenae* and upregulate defense-related gene expression. Pest Manag Sci 75:1310–1316

Chiverton PA (1988) Searching behaviour and cereal aphid consumption by *Bembidion lampros* and *Pterostichus cupreus*, in relation to temperature and prey density. Entomol Exp Appl 47:173–182

Clercq D (2001) Functional response of the predators *Podisus maculiventris* (Say) and *Podisus nigrispinus* (Dallas) (Het., Pentatomidae) to the beet armyworm, *Spodoptera exigua* (Hübner) (Lep., Noctuidae): effect of temperature. J Appl Entomol 125:131–134

Cohen JE, Schoenly K, Heong KL, Justo H, Arida G, Barrion AT et al (1994) A food web approach to evaluating the effect of insecticide spraying on insect pest population dynamics in a Philippine irrigated rice ecosystem. J Appl Ecol 31:747–763

Coyle DR, Nagendra UJ, Taylor MK, Campbell JH, Cunard CE, Joslin (2017) Soil fauna responses to natural disturbances, invasive species, and global climate change: current state of the science and a call to action. Soil Biol Biochem 110: 116-133

Cuong N, Ben P, Phuong L, Chau L, Cohen M (1997) Effect of host plant resistance and insecticide on brown planthopper *Nilaparvata lugens* (Stål) and predator population development in the Mekong Delta, Vietnam. Crop Prot 16:707–715

Dainese M, Martin EA, Aizen M, Albrecht M, Bartomeus I, Bommarco R et al (2019) A global synthesis reveals biodiversity-mediated benefits for crop production. BioRxiv:e554170

Damien M, Tougeron K (2019) Prey–predator phenological mismatch under climate change. Curr Opin Insect Sci 35:60–68

DeLucia EH, Nabity PD, Zavala JA, Berenbaum MR (2012) Climate change: resetting plant-insect interactions. Plant Physiol 160:1677–1685

Dominik C, Seppelt R, Horgan FG, Marquez L, Settele J, Václavík T (2017) Regional-scale effects override the influence of fine-scale landscape heterogeneity on rice arthropod communities. Agric Ecosyst Environ 246:269–278

Dominik C, Seppelt R, Horgan FG, Settele J, Václavík T (2018) Landscape composition, configuration, and trophic interactions shape arthropod communities in rice agroecosystems. J Appl Ecol 55:2461–2472

Dong Z, Hou R, Ouyang Z, Zhang R (2013) Tritrophic interaction influenced by warming and tillage: a field study on winter wheat, aphids and parasitoids. Agric Ecosyst Environ 181:144–148

Dyck V, Thomas B (1979) The brown planthopper problem. In: International Rice Research Institute (ed) Brown planthopper: threat to rice production in Asia. International Rice Research Institute, Los Baños, pp 3–17

Eizaguirre M, López C, Asín L, Albajes R (1994) Thermoperiodism, photoperiodism and sensitive stage in the diapause induction of *Sesamia nonagrioides* (Lepidoptera: Noctuidae). J Insect Physiol 40:113–119

Eizaguirre M, López C, Sans A (2007) Maize phenology influences field diapause induction of *Sesamia nonagrioides* (Lepidoptera: Noctuidae). Bull Entomol Res 92:439–443

EPA (2019) Climate change indicators. Environmental Protection Agency of the United States of America. URL: https://www.epa.gov/climate-indicators/climate-change-indicators-us-and-global-precipitation. Accessed Oct 2019

Fang Y, Liao H, Qian Q, Liu X (2013) Combined effects of temperature and relative humidity on eggs of the rice leaf folder, *Cnaphalocrocis medinalis* (Lepidoptera: Pyralidae). Acta Entomol Sin 56:786–791

Fantinou AA, Kourti AT, Saitanis CJ (2003) Photoperiodic and temperature effects on the intensity of larval diapause in Sesamia nonagrioides. Physiol Entomol 28:82–87

FAO (2019) FAOSTAT crops. Food and Agriculture Organization of the United Nations. URL: http://www.fao.org/faostat/en/#data/QC. Accessed Oct 2019

Farmer AM (1993) The effects of dust on vegetation—a review. Environ Pollut 79:63–75

Farrell J (2016) Corporate funding and ideological polarization about climate change. PNAS 113:92–97

Ferrater JB, Naredo AI, Almazan MLP, de Jong PW, Dicke M, Horgan FG (2015) Varied responses by yeast-like symbionts during virulence adaptation in a monophagous phloem-feeding insect. Arthropod Plant Interact 9:215–224

Folcher L, Bourguet D, Thiéry D, Pélozuelo L, Phalip M, Weissenberger et al (2011) Changes in parasitoid communities over time and space: a historical case study of the maize pest *Ostrinia nubilalis*. PLoS One 6:e25374

Forrest JR (2016) Complex responses of insect phenology to climate change. Curr Opin Insect Sci 17:49–54

Foster SP, Harrington R, Devonshire AL, Denholm I, Devine GJ, Kenward MG et al (2009) Comparative survival of insecticide-susceptible and resistant peach-potato aphids, *Myzus persicae* (Sulzer) (Hemiptera: Aphididae), in low temperature field trials. Bull Entomol Res 86:17–27

Franco AMA, Hill JK, Kitschke C, Collingham YC, Roy DB, Fox R et al (2006) Impacts of climate warming and habitat loss on extinctions at species' low-latitude range boundaries. Glob Chang Biol 12:1545–1553

Gibbard S, Caldeira K, Bala G, Phillips TJ, Wickett M (2005) Climate effects of global land cover change. Geophys Res Lett 32:L23705

Gildor H, Tziperman E (2003) Sea-ice switches and abrupt climate change. Philosophical transactions of the Royal Society of London. Ser A Math Phys Eng Sci 361:1935–1944

Gillyboeuf N, Anglade P, Lavenseau L, Peypelut L (1994) Cold hardiness and overwintering strategy of the pink maize stalk borer, *Sesamia nonagrioides* Lef (Lepidoptera, Noctuidae). Oecologia 99:366–373

Godfray HCJ, Chan MS (1990) How insecticides trigger single-stage outbreaks in tropical pests. Funct Ecol 4:329–337

Gorodkov K (1986) A three-dimensional climatic model of potential distribution area and some of its properties. Entomologicheskoe Obozrenie 65:81–95

Gözel U, Güneş Ç (2013) Effect of entomopathogenic nematode species on the corn stalk borer (*Sesamia cretica* Led. Lepidoptera: Noctuidae) at different temperatures. Turkish J Entomol 37:65–72

Guay J-F, Boudreault S, Michaud D, Cloutier C (2009) Impact of environmental stress on aphid clonal resistance to parasitoids: role of *Hamiltonella defensa* bacterial symbiosis in association with a new facultative symbiont of the pea aphid. J Insect Physiol 55:919–926

Guedes RNC, Cutler GC (2014) Insecticide-induced hormesis and arthropod pest management. Pest Manag Sci 70:690–697

Gurr GM, Lu Z, Zheng X, Xu H, Zhu P, Chen G et al (2016) Multi-country evidence that crop diversification promotes ecological intensification of agriculture. Nat Plants 2:e16014

Hance T, van Baaren J, Vernon P, Boivin G (2006) Impact of extreme temperatures on parasitoids in a climate change perspective. Annu Rev Entomol 52:107–126

Hansen AK, Moran NA (2014) The impact of microbial symbionts on host plant utilization by herbivorous insects. Mol Ecol 23:1473–1496

Hardin MR, Benrey B, Coll M, Lamp WO, Roderick GK, Barbosa P (1995) Arthropod pest resurgence: an overview of potential mechanisms. Crop Prot 14:3–18

Harris RMB, Beaumont LJ, Vance TR, Tozer CR, Remenyi TA, Perkins-Kirkpatrick SE et al (2018) Biological responses to the press and pulse of climate trends and extreme events. Nat Clim Chang 8:579–587

Hatchett JH, Starks KJ, Webster JA (1987) Insect and mite pests of wheat. In: Heyne EG (ed) Wheat and wheat improvement. Agronomy monographs, Wiley, Hoboken, pp 625–675

He L, Asseng S, Zhao G, Wu D, Yang X, Zhuang W et al (2015) Impacts of recent climate warming, cultivar changes, and crop management on winter wheat phenology across the Loess Plateau of China. Agric For Meteorol 200:135–143

Heeb L, Jenner E, Cock MJW (2019) Climate-smart pest management: building resilience of farms and landscapes to changing pest threats. J Pest Sci 92:951–969

Hegland SJ, Nielsen A, Lázaro A, Bjerknes AL, Totland Ø (2009) How does climate warming affect plant-pollinator interactions? Ecol Lett 12:184–195

Heinrichs EA (1994) Biology and management of rice insects. International Rice Research Institute, Los Baños

Herzog T (2009) World greenhouse gas emissions in 2005. World Resources Institute – working paper, Washington DC
Hickling R, Roy DB, Hill JK, Fox R, Thomas CD (2006) The distributions of a wide range of taxonomic groups are expanding polewards. Glob Chang Biol 12:450–455
Hoeppe P (2016) Trends in weather related disasters – consequences for insurers and society. Weather Clim Extremes 11:70–79
Hong J, Lee G-S, Park J-J, Mo H-H, Cho K (2019) Risk map for the range expansion of *Thrips palmi* in Korea under climate change: combining species distribution models with land-use change. J Asia Pac Entomol 22:666–674
Hongyi L, Nan L, Xiaoping X, Baoxin D, Sa Z (2010) Analyzing the influence of meteorology on occurrence and development of wheat aphids in Shandong province. Chin Agric Sci Bull 26:221–225
Hooks CRR, Johnson MW (2004) Using undersown clovers as living mulches: effects on yields, lepidopterous pest infestations, and spider densities in a Hawaiian broccoli agroecosystem. Int J Pest Manag 50:115–120
Hoover JK, Newman JA (2004) Tritrophic interactions in the context of climate change: a model of grasses, cereal aphids and their parasitoids. Glob Chang Biol 10:1197–1208
Horgan FG (2005) Two types of refuge have opposite effects on the size of larval aggregations in a tropical defoliator. Eur J Entomol 102:225–230
Horgan FG (2012) Diversity and defence: plant–herbivore interactions at multiple scales and trophic levels. In: Gurr GM, Wratten SD, Snyder WE, Read DMY (eds) Biodiversity and insect pests: key issues for sustainable management. Wiley, Hoboken, pp 241–258
Horgan FG (2017a) Integrated pest management for sustainable rice cultivation: a holistic approach. In: Sasaki T (ed) Achieving sustainable cultivation of rice—cultivation, pest and disease management. Burleigh Dodds Science, Cambridge, UK, pp 309–342
Horgan FG (2017b) Insect herbivores of rice: their natural regulation and ecologically based management. In: Chauhan BS, Jabran K, Mahajan G (eds) Rice production worldwide. Springer, Cham, pp 279–302
Horgan FG, Ferrater JB (2017) Benefits and potential trade-offs associated with yeast-like symbionts during virulence adaptation in a phloem-feeding planthopper. Entomol Exp Appl 163:112–125
Horgan FG, Myers JH (2004) Interactions between predatory ground beetles, the winter moth and an introduced parasitoid on the lower mainland of British Columbia. Pedobiologia 48:23–35
Horgan FG, Myers JH, Van Meel R (1999) *Cyzenis albicans* (Diptera: Tachinidae) does not prevent the outbreak of winter moth (Lepidoptera: Geometridae) in birch stands and blueberry plots on the lower mainland of British Columbia. Environ Entomol 28:96–107
Horgan FG, Crisol-Martínez E, Almazan MLP, Romena A, Ramal AF, Ferrater JB et al (2016) Susceptibility and tolerance in hybrid and pure-line rice varieties to herbivore attack: biomass partitioning and resource-based compensation in response to damage. Ann Appl Biol 169:200–213
Horgan FG, Ramal AF, Bentur JS, Kumar R, Bhanu KV, Sarao PS et al (2017a) Geographic and research centre origins of rice resistance to Asian planthoppers and leafhoppers: implications for rice breeding and gene deployment. Agronomy 7:e62
Horgan FG, Ramal AF, Villegas JM, Almazan MLP, Bernal CC, Jamoralin A et al (2017b) Ecological engineering with high diversity vegetation patches enhances bird activity and ecosystem services in Philippine rice fields. Reg Environ Chang 17:1355–1367
Horgan FG, Ramal AF, Villegas JM, Jamoralin A, Bernal CC, Perez MO et al (2017c) Effects of bund crops and insecticide treatments on arthropod diversity and herbivore regulation in tropical rice fields. J Appl Entomol 141:587–599
Horgan FG, Bernal CC, Letana S, Naredo AI, Ramp D, Almazan MLP (2018a) Reduced efficiency of tropical flies (Diptera) in the decomposition of snail cadavers following molluscicide poisoning. Appl Soil Ecol 129:61–71

Horgan FG, Kudavidanage EP, Weragodaarachchi A, Ramp D (2018b) Traditional 'maavee' rice production in Sri Lanka: environmental, economic and social pressures revealed through stakeholder interviews. Paddy Water Environ 16:225–241

Horgan FG, Peñalver Cruz A, Bernal CC, Ramal AF, Almazan MLP, Wilby A (2018c) Resistance and tolerance to the brown planthopper, *Nilaparvata lugens* (Stål), in rice infested at different growth stages across a gradient of nitrogen applications. Field Crop Res 217:53–65

Horgan FG, Crisol Martínez E, Stuart AM, Bernal CC, de Cima ME, Almazan MLP et al (2019a) Effects of vegetation strips, fertilizer levels and varietal resistance on the integrated management of arthropod biodiversity in a tropical rice ecosystem. Insects 10:e328

Horgan FG, Srinivasan TS, Crisol-Martínez E, Almazan MLP, Ramal AF, Oliva R et al (2019b) Microbiome responses during virulence adaptation by a phloem-feeding insect to resistant near-isogenic rice lines. Ecol Evol 9:11911–11929

Hu G, Cheng XN, Qi GJ, Wang FY, Lu F, Zhang XX et al (2010a) Rice planting systems, global warming and outbreaks of *Nilaparvata lugens* (Stål). Bull Entomol Res 101:187–199

Hu G, Xie M-C, Lin Z-X, Xin D-Y, Huang C-Y, Chen W et al (2010b) Are outbreaks of *Nilaparvata lugens* (Stål) associated with global warming? Environ Entomol 39:1705–1714

Hu J-H, Wu J-C, Yin J-L, Gu H-N (2010c) Physiology of insecticide-induced stimulation of reproduction in the rice brown planthopper (*Nilaparvata lugens* (Stål)): dynamics of protein in fat body and ovary. Int J Pest Manag 56:23–30

Hu G, Cheng X, Qi G, Wang F, Lu F, Zhang X et al (2011) Rice planting systems, global warming and outbreaks of *Nilaparvata lugens* (Stål). Bull Entomol Res 101:187–199

Hu G, Lu F, Zhai B-P, Lu M-H, Liu W-C, Zhu F et al (2014) Outbreaks of the brown planthopper *Nilaparvata lugens* (Stål) in the Yangtze River Delta: immigration or local reproduction? PLoS One 9:e88973

Hu C, Hou M, Wei G, Shi B, Huang J (2015) Potential overwintering boundary and voltinism changes in the brown planthopper, *Nilaparvata lugens*, in China in response to global warming. Clim Chang 132:337–352

Huang J, Yu H, Guan X, Wang G, Guo R (2015) Accelerated dryland expansion under climate change. Nat Clim Chang 6:e166

Hullé M, d'Acier AC, Bankhead-Dronnet S, Harrington R (2010) Aphids in the face of global changes. C R Biol 333:497–503

Huttunen L, Niemelä P, Peltola H, Heiska S, Rousi M, Kellomäki S (2007) Is a defoliated silver birch seedling able to overcompensate the growth under changing climate? Environ Exp Bot 60:227–238

Ings TC, Montoya JM, Bascompte J, Blüthgen N, Brown L, Dormann CF et al (2009) Review: ecological networks – beyond food webs. J Anim Ecol 78:253–269

Inoue H, Hirao J (1980) Effects of temperature on the transmission of Rice Waika Virus by *Nephotettix cincticeps* Uhler (Hemiptera: Cicadellidae). Appl Entomol Zool 15:433–438

Isichaikul S, Ichikawa T (1993) Relative humidity as an environmental factor determining the microhabitat of the nymphs of the rice brown planthopper, *Nilaparvata lugens* (Stål) (Homoptera: Delphacidae). Res Popul Ecol 35:361–373

Jacobson L, Farstad C (1952) Effect of time of seeding Apex wheat on infestation and sex ratio of the wheat stem sawfly, *Cephus cinctus* Nort.(Hymenoptera: Cephidae). Can Entomol 84:90–92

Jamshidnia A, Kharazi-Pakdel A, Allahyari H, Soleymannejadian E (2010) Functional response of *Telenomus busseolae* (Hym.: Scelionidae) an egg parasitoid of the sugarcane stem borer, *Sesamia nonagrioides* (Lep.: Noctuidae) at different temperatures. Biocontrol Sci Tech 20:631–640

Janzen DH (1967) Why mountain passes are higher in the tropics. Am Nat 101:233–249

Jeffs CT, Lewis OT (2013) Effects of climate warming on host–parasitoid interactions. Ecol Entomol 38:209–218

JISAO (2019) Sahel precipitation index (20-10N, 20W-10E), 1901-2017. Joint Institute for the Study of the Atmosphere and Ocean. URL: http://research.jisao.washington.edu/data/sahel/. Accessed Oct 2019

Jones MG (1978) Development of wheat bulb fly (*Della coarctata* Fall.) larvae and pupae at different temperatures. Entomol Exp Appl 23:288–300

Kaushik C (2011) Incidence of leaf folder [*Cnaphalocrocis mendinalis* (Guenee)] in relation to climatic conditions in the hilly tract of Jalpaiguri, West Bengal. Curr Adv Agric Sci 3:67–68

Kfir R (1997) Competitive displacement of *Busseola fusca* (Lepidoptera: Noctuidae) by *Chilo partellus* (Lepidoptera: Pyralidae). Ann Entomol Soc Am 90:619–624

Khadioli N, Tonnang ZEH, Muchugu E, Ong'amo G, Achia T, Kipchirchir et al (2014) Effect of temperature on the phenology of *Chilo partellus* (Swinhoe) (Lepidoptera, Crambidae); simulation and visualization of the potential future distribution of *C. partellus* in Africa under warmer temperatures through the development of life-table parameters. Bull Entomol Res 104:809–822

Kim H, Lieffering M, Miura S, Kobayashi K, Okada M (2001) Growth and nitrogen uptake of CO_2-enriched rice under field conditions. New Phytol 150:223–229

Kiritani K (1999) Shift of IPM strategy for rice under global warming in temperate areas. In: Journal Department of Zhongshan University (ed) Integrated pest management in rice-based ecosystem. Zhongshan University, China, pp 235-244

Kisimoto R, Sogawa K (1995) Planthopper *Sogatella furcifera* in East Asia: the role of weather and climate. In: Drake VA, Gatehouse AG (eds) Insect migration: tracking resources through space and time. Cambridge University Press, Cambridge, UK, pp 67–92

Klapwijk MJ, Gröbler BC, Ward K, Wheeler D, Lewis OT (2010) Influence of experimental warming and shading on host–parasitoid synchrony. Glob Chang Biol 16:102–112

Knutson TR, Sirutis JJ, Zhao M, Tuleya RE, Bender M, Vecchi GA et al (2015) Global projections of intense tropical cyclone activity for the late twenty-first century from dynamical downscaling of CMIP5/RCP4.5 scenarios. J Clim 28:7203–7224

Kocmánková E, Trnka M, Eitzinger J, Formayer H, Dubrovský M, Semerádová D et al (2010) Estimating the impact of climate change on the occurrence of selected pests in the Central European region. Clim Res 44:95–105

Krishnan S, Chander S (2015) Simulation of climatic change impact on crop-pest interactions: a case study of rice pink stem borer *Sesamia inferens* (Walker). Clim Chang 131:259–272

Krysan JL, Jackson JJ, Lew AC (1984) Field termination of egg diapause in *Diabrotica* with new evidence of extended diapause in *D. barberi* (Coleoptera: Chrysomelidae). Environ Entomol 13:1237–1240

Kuntashula E, Chabala LM, Mulenga BP (2014) Impact of minimum tillage and crop rotation as climate change adaptation strategies on farmer welfare in smallholder farming systems of Zambia. J Sustain Dev 7:95–110

Kvedaras OL, An M, Choi Y-S, Gurr G (2010) Silicon enhances natural enemy attraction and biological control through induced plant defences. Bull Entomol Res 100:367–371

Lacoste C, Nansen C, Thompson S, Moir-Barnetson L, Mian A, McNee M et al (2015) Increased susceptibility to aphids of flowering wheat plants exposed to low temperatures. Environ Entomol 44:610–618

Larsson S (1989) Stressful times for the plant stress: insect performance hypothesis. Oikos 56:277–283

Levine E, Oloumi-Sadeghi H, Ellis CR (1992a) Thermal requirements, hatching patterns, and prolonged diapause in western corn rootworm (Coleoptera: Chrysomelidae) eggs. J Econ Entomol 85:2425–2432

Levine E, Oloumi-Sadeghi H, Fisher JR (1992b) Discovery of multiyear diapause in Illinois and South Dakota northern corn rootworm (Coleoptera: Chrysomelidae) eggs and incidence of the prolonged diapause trait in Illinois. J Econ Entomol 85:262–267

Li B, Guo Q, Meng L (2013) Effects of elevated CO_2 concentration on development, reproduction and food utilization of the *Cnaphalocrocis medinalis* Guenée (Lepidoptera: Pyralidae). Sci Agric Sin 46:4464–4470

Lobell DB, Schlenker W, Costa-Roberts J (2011) Climate trends and global crop production since 1980. Science 333:616–620

Loo YY, Billa L, Singh A (2015) Effect of climate change on seasonal monsoon in Asia and its impact on the variability of monsoon rainfall in Southeast Asia. Geosci Front 6:817–823

LoPresti EF, Karban R (2016) Chewing sandpaper: grit, plant apparency and plant defense in sand-entrapping plants. Ecology 97:826–833

Lu X-F, Huo Z-G, Shen S-h, Huang D-P, Wang L, Xiao J-J et al (2012) Effects of climate warming on the northern distribution boundary of brown planthopper (*Nilaparvata lugens* (Stål)) overwintering in China. Chin J Ecol 31:1977–1983

Lu K, Chen X, Liu W, Zhou Q (2016) Identification of a heat shock protein 90 gene involved in resistance to temperature stress in two wing-morphs of *Nilaparvata lugens* (Stål). Comp Biochem Physiol A Mol Integr Physiol 197:1–8

Lucas PW, van Casteren A, Al-Fadhalah K, Almusallam AS, Henry AG, Michael S et al (2014) The role of dust, grit and phytoliths in tooth wear. Ann Zool Fenn 51:143–152

Luo J, Liu Y, Gong Y, Cheng X, Fu Q, Hu G (2013) Investigation of the overwintering of three species of rice pest, *Nilaparvata lugens*, *Sogatella furcifera* and *Cnaphalocrocis medinalis* in China. Chin J Appl Entomol 50:253–260

Lüthi D, Le Floch M, Bereiter B, Blunier T, Barnola J-M, Siegenthaler U et al (2008) High-resolution carbon dioxide concentration record 650,000–800,000 years before present. Nature 453:379–382

Ma G, Ma C-S (2012) Climate warming may increase aphids' dropping probabilities in response to high temperatures. J Insect Physiol 58:1456–1462

Ma M-Y, Peng Z-P, He Y (2012) Effects of temperature on functional response of *Anagrus nilaparvatae* Pang et Wang (Hymenoptera: Mymaridae) on the eggs of whitebacked planthopper, *Sogatella furcifera* Horváth and brown planthopper, *Nilaparvata lugens* Stål. J Integr Agric 11:1313–1320

Mann ME, Emanuel KA (2006) Atlantic hurricane trends linked to climate change. Trans Am Geophys Union 87:233–241

Manns HR, Maxwell CD, Emery RJN (2007) The effect of ground cover or initial organic carbon on soil fungi, aggregation, moisture and organic carbon in one season with oat (*Avena sativa*) plots. Soil Tillage Res 96:83–94

Mantyka-Pringle CS, Visconti P, Di Marco M, Martin TG, Rondinini C, Rhodes JR (2015) Climate change modifies risk of global biodiversity loss due to land-cover change. Biol Conserv 187:103–111

Matsumura M, Takeuchi H, Satoh M, Sanada-Morimura S, Otuka A, Watanabe T et al (2008) Species-specific insecticide resistance to imidacloprid and fipronil in the rice planthoppers *Nilaparvata lugens* and *Sogatella furcifera* in East and South-East Asia. Pest Manag Sci 64:1115–1121

Meisner MH, Harmon JP, Ives AR (2014) Temperature effects on long-term population dynamics in a parasitoid–host system. Ecol Monogr 84:457–476

Menéndez R, González-Megías A, Jay-Robert P, Marquéz-Ferrando R (2014) Climate change and elevational range shifts: evidence from dung beetles in two European mountain ranges. Glob Ecol Biogeogr 23:646–657

Mia A-M, Iwahashi O (1999) Seasonal changes in infestation level of sugarcane by the pink borer, *Sesamia inferens* (Lepidoptera: Noctuidae), in relation to a parasitoid, *Cotesia flavipes* (Hymenoptera: Braconidae), on Okinawa Island. Appl Entomol Zool 34:429–434

Mochida OM, Joshi RC, Litsinger JA (1987) Climatic factors affecting the occurrence of insect pests. Weather and rice: proceedings of the international workshop on the impact of weather parameters on growth and yield of rice. International Rice Research Institute, Los Baños, pp 149–164

Moreau G (2006) Past and present outbreaks of the balsam fir sawfly in western Newfoundland: an analytical review. For Ecol Manag 221:215–219

Morishita M (1992) A possible relationship between outbreaks of planthoppers, *Nilaparvata lugens* Stål and *Sogatella furcifera* Horváth (Hemiptera: Delphacidae) in Japan and the El Niño phenomenon. Appl Entomol Zool 27:297–299

Musser FR, Shelton AM (2005) The influence of post-exposure temperature on the toxicity of insecticides to *Ostrinia nubilalis* (Lepidoptera: Crambidae). Pest Manag Sci 61:508–510

Mutamiswa R, Chidawanyika F, Nyamukondiwa C (2018) Thermal plasticity potentially mediates the interaction between host *Chilo partellus* Swinhoe (Lepidoptera: Crambidae) and endoparasitoid *Cotesia flavipes* Cameron (Hymenoptera: Braconidae) in rapidly changing environments. Pest Manag Sci 74:1335–1345

Myers JH (2000) Population fluctuations of the western tent caterpillar in southwestern British Columbia. Popul Ecol 42:231–241

NASA (2019) Global vital signs: vital signs of the planet. North Americean Space Agency. URL: https://climate.nasa.gov/. Accessed Oct 2019

Nascimento AM, Assis FA, Moraes JC, Souza BHS (2018) Silicon application promotes rice growth and negatively affects development of *Spodoptera frugiperda* (J. E. Smith). J Appl Entomol 142:241–249

Nelson GC, Rosegrant MW, Koo J, Robertson R, Sulser T, Zhu T et al (2009) Climate change: impact on agriculture and costs of adaptation. International Food Policy Research Institute, Washington DC

Newman JA (2004) Climate change and cereal aphids: the relative effects of increasing CO_2 and temperature on aphid population dynamics. Glob Chang Biol 10:5–15

Nicholls RJ, Cazenave A (2010) Sea-level rise and its impact on coastal zones. Science 328:1517–1520

Norris JR, Allen RJ, Evan AT, Zelinka MD, O'Dell CW, Klein SA (2016) Evidence for climate change in the satellite cloud record. Nature 536:72–75

Ntiri ES, Calatayud P-A, Van Den Berg J, Schulthess F, Le Ru BP (2016) Influence of temperature on intra- and interspecific resource utilization within a community of lepidopteran maize stemborers. PLoS One 11:e0148735

Olesen JE, Trnka M, Kersebaum KC, Skjelvåg AO, Seguin B, Peltonen-Sainio P et al (2011) Impacts and adaptation of European crop production systems to climate change. Eur J Agron 34:96–112

Ortega A (1987) Insect pests of maize: a guide for field identification. Centro Internacional de Mejoramiento de Maiz y Trigo, Mexico

Otuka A, Matsumura M, Watanabe T, Van Dinh T (2008) A migration analysis for rice planthoppers, *Sogatella furcifera* (Horváth) and *Nilaparvata lugens* (Stål) (Homoptera: Delphacidae), emigrating from northern Vietnam from April to May. Appl Entomol Zool 43:527–534

Otuka A, Huang S-H, Sanada-Morimura S, Matsumura M (2012) Migration analysis of *Nilaparvata lugens* (Hemiptera: Delphacidae) from the Philippines to Taiwan under typhoon-induced windy conditions. Appl Entomol Zool 47:263–271

Overpeck J, Hughen K, Hardy D, Bradley R, Case R, Douglas M et al (1997) Arctic environmental change of the last four centuries. Science 278:1251–1256

Ôya S (1979) Survival tests on submerged green rice leafhopper: *Nephotettix cincticeps* Uhler, in relation to winter mortality under snow cover in the Hokuriku District. Appl Entomol Zool 14:319–325

Pachauri RK, Myles RA, Barros VR, Broome J, Cramer W, Christ R et al (2014) Climate change 2014: synthesis report, Fifth assessment report of the Intergovernmental Panel on Climate Change 151. Intergovernmental Panel on Climate Change, Geneva

Pandi GGP, Chander S, Singh MP, Pathak H (2018) Impact of elevated CO_2 and temperature on brown planthopper population in rice ecosystem. Proc Nat Acad Sci India Sect B Biol Sci 88:57–64

Parker DR, Page AL (1994) Vegetation management strategies for remediation of selenium contaminated soils. Marcel Dekker, New York

Patel H, Kadu R, Landge S (2011) Study on seasonal incidence of rice leaf folders (*Cnaphalocrocis medinalis* Guen. and *Pelopidas mathias* Fb.) of paddy and its correlation with weather parameters. Int J Plant Prot 4:175–180

Pecl GT, Araújo MB, Bell JD, Blanchard J, Bonebrake TC, Chen I-C et al (2017) Biodiversity redistribution under climate change: impacts on ecosystems and human well-being. Science 355:eaai9214

Pelletier Y, Horgan FG, Pompon J (2013) Potato resistance against insect herbivores: resources and opportunities. In: Alyokhin A, Vincent C, Giordanengo P (eds) Insect pests of potato: global perspectives on biology and management. Academic Press, Cambridge, pp 439–462

Pérez-Rodríguez J, Shortall CR, Bell JR (2015) Large-scale migration synchrony between parasitoids and their host. Ecol Entomol 40:654–659

Piyaphongkul J, Pritchard J, Bale J (2014) Effects of acclimation on the thermal tolerance of the brown planthopper *Nilaparvata lugens* (Stål). Agric For Entomol 16:174–183

Prasannakumar NR, Chander S, Pal M (2012) Assessment of impact of climate change with reference to elevated CO_2 on rice brown planthopper, *Nilaparvata lugens* (Stal.) and crop yield. Curr Sci 103:1201–1205

Preetha G, Stanley J, Suresh S, Samiyappan R (2010) Risk assessment of insecticides used in rice on miridbug, *Cyrtorhinus lividipennis* Reuter, the important predator of brown planthopper, *Nilaparvata lugens* (Stål.). Chemosphere 80:498–503

Punzo F (1993) Detoxification enzymes and the effects of temperature on the toxicity of pyrethroids to the fall armyworm, *Spodoptera frugiperda* (Lepidoptera: Noctuidae). Compar Biochem Physiol Part C: Compar Pharmacol 105:155–158

Qian L, Liu X, Huang Z, Wang L, Zhang Y, Gao Y et al (2019) Elevated CO_2 enhances the host resistance against the western flower thrips, *Frankliniella occidentalis*, through increased callose deposition. J Pest Sci (online)

Qiu J, Tang H, Frolking S, Boles S, Li C, Xiao X et al (2003) Mapping single-, double-, and triple-crop agriculture in China at 0.5° × 0.5° by combining county-scale census data with a remote sensing-derived land cover map. Geocarto Int 18:3–13

Raga IN, Ito K, Matsui M, Okada M (1988) Effects of temperature on adult longevity, fertility, and rate of transovarial passage of Rice Stripe Virus in the small brown planthopper, *Laodelphax striatellus* Fallen (Homoptera: Delphacidae). Appl Entomol Zool 23:67–75

Ramirez-Cabral NYZ, Kumar L, Shabani F (2017) Future climate scenarios project a decrease in the risk of fall armyworm outbreaks. J Agric Sci 155:1219–1238

Ray DK, West PC, Clark M, Gerber JS, Prishchepov AV, Chatterjee S (2019) Climate change has likely already affected global food production. PLoS One 14:e0217148

Reidsma P, Ewert F, Lansink AO, Leemans R (2010) Adaptation to climate change and climate variability in European agriculture: the importance of farm level responses. Eur J Agron 32:91–102

Riley J, Reynolds D, Smith A, Rosenberg L, Xia-nian C, Xiao-xi Z et al (1994) Observations on the autumn migration of *Nilaparvata lugens* (Homoptera: Delphacidae) and other pests in east central China. Bull Entomol Res 84:389–402

Riley J, Reynolds D, Smith A, Edwards A, Zhang X-X, Cheng X-N et al (1995) Observations of the autumn migration of the rice leaf roller *Cnaphalocrocis medinalis* (Lepidoptera: Pyralidae) and other moths in eastern China. Bull Entomol Res 85:397–414

Roberts DWA, Tyrrell C (1961) Sawfly resistance in wheat: IV. Some effects of light intensity on resistance. Can J Plant Sci 41: 457-465

Roitberg BD, Myers JH (2012) Behavioural and physiological adaptations of pea aphids (Homoptera: Aphididae) to high ground temperatures and predator disturbance. Can Entomol 111:515–519

Rosenberg L, Magor J (1987) Predicting windborne displacements of the brown planthopper, *Nilaparvata lugens* from synoptic weather data. 1. Long-distance displacements in the north-east monsoon. J Anim Ecol 56:39–51

Rosenzweig C, Iglesias A, Yang XB, Epstein PR, Chivian E (2001) Climate change and extreme weather events; implications for food production, plant diseases, and pests. Glob Change Human Health 2:90–104

Rozema J, Flowers T (2008) Crops for a salinized world. Science 322:1478–1480

Runjie Z, Heong KL, Domingo IT (1996) Relationship between temperature and functional response in *Cardiochiles philippinensis* (Hymenoptera: Braconidae), a larval parasitoid of *Cnaphalocrocis medinalis* (Lepidoptera: Pyralidae). Environ Entomol 25:1321–1324

Russell JA, Moran NA (2005) Costs and benefits of symbiont infection in aphids: variation among symbionts and across temperatures. Proc R Soc B Biol Sci 273:603–610

Sakamoto T, Van Nguyen N, Ohno H, Ishitsuka N, Yokozawa M (2006) Spatio–temporal distribution of rice phenology and cropping systems in the Mekong Delta with special reference to the seasonal water flow of the Mekong and Bassac rivers. Remote Sens Environ 100:1–16

Salminen SO, Richmond DS, Grewal SK, Grewal PS (2005) Influence of temperature on alkaloid levels and fall armyworm performance in endophytic tall fescue and perennial ryegrass. Entomol Exp Appl 115:417–426

Sánchez-Bayo F, Wyckhuys KA (2019) Worldwide decline of the entomofauna: a review of its drivers. Biol Conserv 232:8–27

Schmidt MH, Thewes U, Thies C, Tscharntke T (2004) Aphid suppression by natural enemies in mulched cereals. Entomol Exp Appl 113:87–93

Schoenly KG, Cohen JE, Heong KL, Arida GS, Barrion AT, Litsinger JA (1996) Quantifying the impact of insecticides on food web structure of rice-arthropod populations in a Philippine farmer's irrigated field: a case study. In: Polis GA, Winemiller KO (eds) Food webs. Springer, Boston, pp 343–351

Schuldiner-Harpaz T, Coll M (2013) Effects of global warming on predatory bugs supported by data across geographic and seasonal climatic gradients. PLoS One 8:e66622

Schweissing F, Wilde G (1979) Temperature and plant nutrient effects on resistance of seedling sorghum to the greenhug. J Econ Entomol 72:20–23

Sedlock JL, Stuart AM, Horgan FG, Hadi B, Como Jacobson A, Alviola PA et al (2019) Local-scale bat guild activity differs with rice growth stage at ground level in the Philippines. Diversity 11:e148

Settle WH, Ariawan H, Astuti ET, Cahyana W, Hakim AL, Hindayana D et al (1996) Managing tropical rice pests through conservation of generalist natural enemies and alternative prey. Ecology 77:1975–1988

Shakoor A (2014) Silicon biomineralisation in plants: a tool to adapt to global climate change. J Res Biol Sci 1:1–3

Shi B-K, Huang J-L, Hu C-X, Hou M-L (2014) Interactive effects of elevated CO_2 and temperature on rice planthopper, *Nilaparvata lugens*. J Integr Agric 13:1520–1529

Singh A (2017) Farmers' perception and adoption of abiotic stress tolerant rice varieties in rain-fed lowlands of North Eastern Uttar Pradesh. Ind J Ext Edu 53:32–36

Singh J, Singh JP (1995) Land degradation and economic sustainability. Ecol Econ 15:77–86

Song Y, Heong K (1997) Changes in searching responses with temperature of *Cyrtorhinus lividipennis* Reuter (Hemiptera: Miridae) on the eggs of the brown planthopper, *Nilaparvata lugens* (Stål.) (Homoptera: Delphacidae). Popul Ecol 39:201–206

Sparks TC, Shour MH, Wellemeyer EG (1982) Temperature-toxicity relationships of pyrethroids on three lepidopterans. J Econ Entomol 75:643–646

Srinivasan TS, Almazan MLP, Bernal CC, Ramal AF, Subbarayalu MK, Horgan FG (2016) Interactions between nymphs of *Nilaparvata lugens* and *Sogatella furcifera* (Hemiptera: Delphacidae) on resistant and susceptible rice varieties. Appl Entomol Zool 51:81–90

Stacey DA, Fellowes MDE (2007) Influence of temperature on pea aphid *Acyrthosiphon pisum* (Hemiptera: Aphididae) resistance to natural enemy attack. Bull Entomol Res 92:351–357

Stamp NE, Osier TL (1998) Response of five insect herbivores to multiple allelochemicals under fluctuating temperatures. Entomol Exp Appl 88:81–96

Stireman JO, Dyer LA, Janzen DH, Singer MS, Lill JT, Marquis RJ et al (2005) Climatic unpredictability and parasitism of caterpillars: implications of global warming. PNAS 102:e17384

Sujithra M, Chander S (2013) Simulation of rice brown planthopper, *Nilaparvata lugens* (Stål.) population and crop-pest interactions to assess climate change impact. Clim Chang 121:331–347

Susanti E, Ramadhani F, June T, Amien L (2010) Utilization of climate information for development of early warning system for brown plant hopper attack on rice. Indonesian J Agric 3:17–25
Svobodová E, Trnka M, Dubrovský M, Semerádová D, Eitzinger J, Štěpánek P et al (2014) Determination of areas with the most significant shift in persistence of pests in Europe under climate change. Pest Manag Sci 70:708–715
Takagi H, Esteban M (2016) Statistics of tropical cyclone landfalls in the Philippines: unusual characteristics of 2013 Typhoon Haiyan. Nat Hazards 80:211–222
Tang J, Zhu W, Kookana R, Katayama A (2013) Characteristics of biochar and its application in remediation of contaminated soil. J Biosci Bioeng 116:653–659
Tegen I, Werner M, Harrison SP, Kohfeld KE (2004) Relative importance of climate and land use in determining present and future global soil dust emission. Geophys Res Lett 31:L05105
Thomson LJ, Macfadyen S, Hoffmann AA (2010) Predicting the effects of climate change on natural enemies of agricultural pests. Biol Control 52:296–306
Tougeron K, Tena A (2019) Hyperparasitoids as new targets in biological control in a global change context. Biol Control 130:164–171
Tougeron K, Brodeur J, Le Lann C, van Baaren J (2019) How climate change affects the seasonal ecology of insect parasitoids. Ecol Entomol (online)
Tougou D, Musolin DL, Fujisaki K (2009) Some like it hot! Rapid climate change promotes changes in distribution ranges of *Nezara viridula* and *Nezara antennata* in Japan. Entomol Exp Appl 130:249–258
Uphoff N (1999) Agroecological implications of the system of rice intensification (SRI) in Madagascar. Environ Dev Sustain 1:297–313
Van Dyck H, Bonte D, Puls R, Gotthard K, Maes D (2015) The lost generation hypothesis: could climate change drive ectotherms into a developmental trap? Oikos 124:54–61
Vu Q, Ramal AF, Villegas JM, Jamoralin A, Bernal CC, Pasang JM et al (2018) Enhancing the parasitism of insect herbivores through diversification of habitat in Philippine rice fields. Paddy Water Environ 16:379–390
Wang B, Huang F, Wu Z, Yang J, Fu X, Kikuchi K (2009a) Multi-scale climate variability of the South China Sea monsoon: a review. Dyn Atmos Oceans 47:15–37
Wang F, Hu G, Chen X, Shen H, Luo S, Xin D et al (2009b) Analysis on the causes of recent outbreaks of *Cnaphalocrocis medinalis* in Nanning, China. Chin J Rice Sci 23:537–545
Wang B-j, Xu H-x, Zheng X-s, Qiang F, Lu Z-x (2010) High temperature modifies resistance performances of rice varieties to brown planthopper, *Nilaparvata lugens* (Stål). Rice Sci 17:334–338
WMO/FAO (2016) Weather and desert locusts. World Meteorological Organization and Food and Agriculture Organization of the United Nations, Rome
Wu G, Chen FJ, Ge F (2006) Response of multiple generations of cotton bollworm *Helicoverpa armigera* Hübner, feeding on spring wheat, to elevated CO_2. J Appl Entomol 130:2–9
Wu L-H, Hill MP, Thomson LJ, Hoffmann AA (2018) Assessing the current and future biological control potential of *Trichogramma ostriniae* on its hosts *Ostrinia furnacalis* and *Ostrinia nubilalis*. Pest Manag Sci 74:1513–1523
Wu Y, Gong Z, Bebber DP, Miao J, Zhao Z, Jiang Y et al (2019) Phenological matching drives wheat pest range shift under climate change. bioRxiv e614743
Wu J, Ge L, Liu F, Song Q, Stanley D (2020) Pesticide-induced planthopper population resurgence in rice cropping systems. Annu Rev Entomol 65:409–429
Xu H, Xie H, Wu S, Wang Z, He K (2019) Effects of elevated CO_2 and increased N fertilization on plant secondary metabolites and chewing insect fitness. Front Plant Sci 10:e739
Yamano T, Malabayabas M, Gumma M (2013) Adoption, yield, and ex ante impact analysis of Swarna-Sub 1 in Eastern India. STRASA Econ Briefs 2:3–10
Yin J, Sun Y, Wu G, Ge F (2010) Effects of elevated CO_2 associated with maize on multiple generations of the cotton bollworm, *Helicoverpa armigera*. Entomol Exp Appl 136:12–20

Yuan X-H, Song L-W, Zhang J-J, Zang L-S, Zhu L, Ruan C-C et al (2012) Performance of four Chinese *Trichogramma* species as biocontrol agents of the rice striped stem borer, *Chilo suppressalis*, under various temperature and humidity regimes. J Pest Sci 85:497–504

Yukawa J, Kiritani K, Kawasawa T, Higashiura Y, Sawamura N, Nakada K et al (2009) Northward range expansion by *Nezara viridula* (Hemiptera: Pentatomidae) in Shikoku and Chugoku Districts, Japan, possibly due to global warming. Appl Entomol Zool 44:429–437

Zhang Z-T, Li J, Sun W (2013) Influences of droughts and floods disasters on the population dynamics of Asian corn borer, *Ostrinia furnacalis* (Guenee). J Maize Sci 28

Zhao Z, Sandhu HS, Ouyang F, Ge F (2016) Landscape changes have greater effects than climate changes on six insect pests in China. Sci China Life Sci 59:627–633

Zhou C, Yang H, Yang H, Wang Z, Long G-Y, Jin D-C (2019) Effects of sublethal concentrations of deltamethrin on fitness of white-backed planthopper, *Sogatella furcifera* (Horváth). Int J Pest Manag 65:165–170

Crop Protection Under Drought Stress

Ram L. Ray, Peter A. Y. Ampim, and Ming Gao

Introduction

Climate change and its effect on the variability of weather patterns have a significant impact on agricultural practices, the availability of natural resources, and the nature of the environment. According to the National Climate Assessment (NCA), climate change will continue to have a significant impact on crop production and agricultural practices over the next few decades and possibly beyond. Because of these issues, climate change will significantly impact global food security and terrestrial ecosystems. The complexity of these problems is shown by the increase in the frequency and intensity of droughts in some regions around the world and the increase in the intensity of heavy precipitation events on a global scale (IPCC 2019). The Intergovernmental Panel on Climate Change (IPCC) (2019) has predicted a temperature rise of 1.5 °C between 2030 and 2052, plus a significant change in precipitation patterns, which, together with a greater frequency of extreme weather events, will significantly affect agricultural production. These findings provide strong evidence that human-driven emission of greenhouse gases is causing climate change risks, which should not be ignored. In this respect, it is important to understand that the global mean land surface air temperature is increasing faster than the global mean surface temperature (combined land surface and sea surface temperature) (Fig. 1).

Climate variables, such as temperature and precipitation, have direct impact on crop production because they contribute to crop growth, health, and yield, thus affecting cropping system efficiency over time (Ray et al. 2018; Howden et al. 2007; Kang et al. 2009; Lehmann 2013; Paudel et al. 2014; Liang et al. 2017). In the future, climate extremes are expected to increase due to the effects of climate

R. L. Ray (✉) · P. A. Y. Ampim · M. Gao
Cooperative Agricultural Research Center, College of Agriculture and Human Sciences, Prairie View A&M University, Prairie View, TX, USA
e-mail: raray@pvamu.edu; paampim@pvamu.edu; migao@pvamu.edu

K. Jabran et al. (eds.), *Crop Protection Under Changing Climate*,
https://doi.org/10.1007/978-3-030-46111-9_6

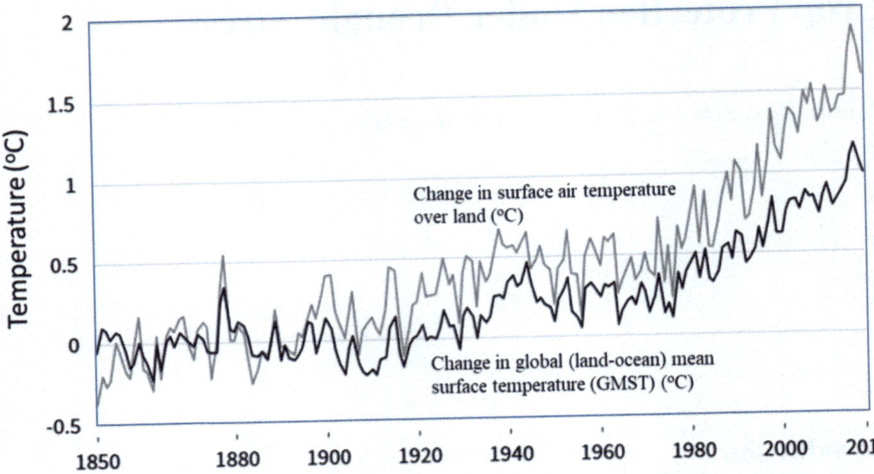

Fig. 1 Change in air and mean temperatures from 1850 to 1900. (Modified from IPCC 2019)

change, which may significantly increase the negative impacts on crop production (Troy et al. 2015). Given this scenario, it is remarkable that numerous researchers have studied the effects of climate change on agriculture. However, past studies have not focused on adaptive changes to improve cropping practices to manage the impact of drought on crop production (Troy et al. 2015).

Water stress resulting from drought is known to reduce crop production because of its negative impacts on plant growth (Karl et al. 2009). Plants, including crops, are naturally subjected to a variety of abiotic stresses such as drought, salinity, heat, and other factors in their life cycle (Manzoor et al. 2016; Hussain et al. 2018; Tandzi et al. 2019; Nabi et al. 2019) and are equipped with different resistance mechanisms for such stresses, the effectiveness of which vary from species to species and even within species (Manzoor et al. 2016; Jaleel et al. 2009). Particularly, in the context of drought, some crops have high drought tolerance capacity (e.g., pomegranate, sorghum, cassava, millet, sweet potato), while others have low tolerance capacity (e.g., sugarcane, banana, citrus, cotton, rice). Mechanism of drought tolerance in the plant is a complex phenomenon as interactions between stress factors and different molecular, biochemical, and physiological factors affect crop growth and development (Jaleel et al. 2009; Razmjoo et al. 2008). Therefore, it is important to understand the impact of water stress and drought on crop growth and its development, physiological process, morphology, and yields and available genetic and agronomic tools for crop protection from drought.

Drought stress is a critical limiting factor at the initial stage of plant's physical growth and development, determining plant height, stem size, number of and size of leaves, flower and fruit production, root size and distribution, and seed development. Moreover, drought stress causes a change in the physical environment, which subsequently affects physiological and biochemical processes in plants (Silva et al. 2009; Fathi and Tari 2016). Water stress causes negative effects on the overall growth and development of crops, resulting in a significant reduction in crop production, which will contribute to a reduction of global food supplies (Lesk et al. 2016). However, proper strategies for drought mitigation combined with the best agricultural management practices can reduce the impact of climate extremes on crop production under changing climate effects.

These "best management practices" that contribute to drought adaptation due to climate change, and which support mitigation processes, include appropriate agronomic and genetic tools for crop protection under drought. For example, during drought events, it is important to have planned strategies on how best to (i) utilize available water resources, (ii) scale back on acreage to be planted, (iii) select early maturing and drought-tolerant crop varieties, (iv) select the most effective irrigation practices, and (v) use reduced tillage practices. These strategies are suggested because it has been observed that sustainable agricultural management practices are not widely adopted due to lack of access to resources, knowledge, and practical experiences. In addition, it is necessary to continue our efforts on selecting improved varieties of all crops for better yield and higher quality and expanded cultivation environment to enhance their drought tolerance. It is possible to enhance the drought tolerance limit of a crop by introducing foreign genetic materials that confer added drought tolerance through genetic transformation. This is a recent biotechnological approach that shows much promise (Rejeb et al. 2016).

The aim of this chapter is to provide a critical and comprehensive review of recent studies related to the impact of climate extremes, such as drought, on crop physiology, crop morphology, and crop yields. It will also investigate issues of global food security and available genetic and agronomic tools in addressing drought stress and the protection of crops under drought conditions. Furthermore, this chapter is focused on adaptation strategies to mitigate the effects of drought and to augment crop management for sustainable and climate-smart agriculture. This assessment will provide a technical review of climate-smart agriculture, which may assist farmers and growers to better understand crop needs under changing climate conditions.

Effects of Drought on the Physiological Processes of Crop Plants

Plant growth, physiology, and reproduction are negatively impacted during severe droughts (Fig. 2), which causes substantial decline in crop yields (Yordanov et al. 2000, 2003; Farooq et al. 2009). As shown in Fig. 2, cell elongation in higher plants under drought stress is inhibited by reduced turgor pressure. Reduction in water uptake caused a reduction in tissue water content. Turgor is lost due to a lack of water. Similarly, drought stress also limits the photo assimilation and metabolites which are essential for cell division. Moreover, under drought stress, impaired mitosis, cell elongation, and expansion result in reduced crop growth, leave parameters such as leaf length, and leaf area index (Farooq et al. 2009).

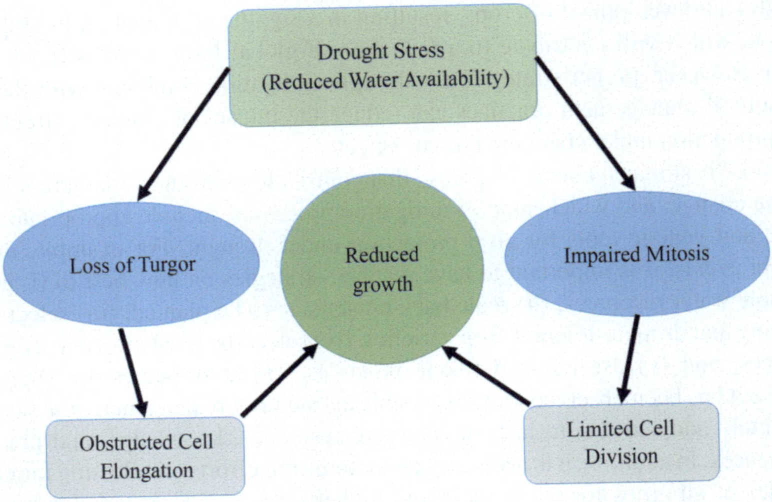

Fig. 2 Description of possible mechanisms of growth reduction under drought stress. (Modified from Farooq et al. 2009)

Harris et al. (2002), in their review, noted that the foremost effect of drought is weak growth and poor stand establishment, and other studies have also indicated that drought has a significant impact on germination as well as seedling stand (Kaya et al. 2006; Farooq et al. 2009). As shown in Fig. 2, plant growth occurs through cell enlargement and cell division, which involves genetic, ecological, physiological, and morphological processes and their complex interactions (Fahad et al. 2017). The quality and quantity of plant growth depend on these processes, and it is important to note that they are significantly affected by water deficit. Under critical water deficiency, cell elongation of higher plants can be reduced by water flow interruption from the xylem to the surrounding elongating cells (Nonami 1998). Plants need nutrients and sufficient water throughout their growth period in order to allow maximum production (Silva et al. 2013), and thus a reduction in water content in the soil intimately affects plant growth and development. As a result of reductions in soil moisture, changes in the physical environment occur, which subsequently affect physiological and biochemical processes in plants (Sarker et al. 2005; Sircelj et al. 2005; Silva et al. 2009; Fathi and Tari 2016). Water is also essential for photosynthesis, respiration, and other physiological and biochemical processes of plant growth (Farooq et al. 2009). Therefore, when there is unavailability or shortage of water, changes inevitably occur in all aspects of plant growth and development.

Physiological parameters include net CO_2 assimilation rate (Pn), transpiration rate (T), stomatal conductance (*gs*), chlorophyll content, leaf water potential (lwp), and water use efficiency (WUE). However, the process of photosynthesis includes all of the physiological parameters in the crop growth cycle, which are also termed "photosynthetic parameters." A major effect of drought is the reduction of photosynthesis within a plant, which arises from the changes in net CO_2 assimilation rate, transpiration rate, stomatal conductance, chlorophyll content, leaf water potential, water use efficiency, and other factors (Athar and Ashraf 2005). For example, under drought stress, Pn, gs, WUE, T, adenosine triphosphate (ATP), photochemical quenching, and rubisco protein activity are decreased. Conversely, non-photochemical quenching is increased, which ultimately affects photosynthesis and plant growth. In addition, the earliest response to drought is stomatal closure, which decreases photosynthesis but protects the plant from extensive water loss, which might cause cell dehydration and death (Athar and Ashraf 2005; Farooq et al. 2009).

After the stomatal closure, CO_2 levels inside the leaf and transpiration rates start to decrease, which causes an increase in heat (Yokota et al. 2002). In the past, researchers also found that the stomatal response to drought is more closely linked to soil moisture than the leaf water (Farooq et al. 2009). In addition, the rate of stomatal closure is proportional to the rate of increase in drought stress. However, physiological parameters are not controlled by soil moisture availability alone; rather, they are also impacted by other complex interactions among intrinsic and extrinsic factors such as plant traits, phenological strategies, and hydro-climatic drivers (Vico et al. 2017; Farooq et al. 2009).

Plants can respond, adapt, and survive under drought stress by using various drought resistance mechanisms linked to biochemical, morphological, and physiological parameters. Since drought stress affects the plant's water balance and its effects at the cellular, tissue, and organ levels, proper physiological, molecular, and morphological mechanisms are important for drought mitigation. For example, plants may control/limit drought stress by reducing the growing period and maintaining high tissue water potential either by reducing water depletion from plants or improving plant water uptake (Farooq et al. 2009). Osmotic adjustment, osmoprotection, antioxidation, and a scavenging defense system are the essential bases responsible for drought resistance. According to Farooq et al. (2009), cell and tissue water conservation, an antioxidant defense system, cell membrane stability, aquaporins, and stress proteins are important mechanisms for the drought resistance. Moreover, drought stress can also be managed by the production of appropriate genotypes, seed priming, plant growth regulators, and the use of silicon, osmoprotectants, and others.

Effects of Drought on Morphology of Crop Plants

Drought, among other environmental factors, is an important environmental stress that weakens plant growth and development (Shao et al. 2008; Tátrai et al. 2016). Drought stress occurs in plants either when the water supply to roots becomes limited or when evapotranspiration of water from plants becomes very high (Anjum et al. 2017). Plant growth and developmental processes affected by drought include alterations in germination, plant height, stem size, number of leaves and their sizes, flower and fruit productions, root size and distribution, seed development, yield, and quality (Anjum et al. 2017; Jaleel et al. 2007).

The effect of drought on the morphology of plants includes a decrease in stem length, stem diameter, volume of leaves, leaf size, and leaf area and a reduction in plant height (Riaz et al. 2013). For example, Specht et al. (2001) found a reduction in stem diameter of soybean plants, Wu et al. (2008) found a reduction in the height of citrus seedlings under water deficit conditions, and Tangu (2014) found a significant reduction in volume of leaves, leaf size, and leaf area of olive plants under drought stress. Moisture stress induces plant structural changes, which are all critical in responding to drought stress, and it has been commented that a deep rooting system is a "drought avoidance strategy" (Hund et al. 2009). Effective plant drought tolerance includes changes at the tissue and molecular levels and the exposure of the plant to a single occurrence or combination of these basic changes, which determines the ability of the plant to sustain itself under low water content.

While plant growth is supported by mitosis, cell elongation, and differentiation, drought stress can impair mitosis and cell elongation, resulting in poor growth because water is a major component of plant cells and facilitates germination and growth processes. Also, plant growth includes an increase in volume, size, or weight and enhances the process of seed germination, which requires healthy soil, adequate sunlight, and sufficient water. In addition, favorable climatic and hydrologic parameters (e.g., temperatures and soil moisture) also play a significant role in enhancing the process of plant growth (Farooq et al. 2009). Several studies have shown how the negative impacts of drought and heat stress substantially affect seed yields by reducing seed size and number (Fahad et al. 2017; Kaya et al. 2006; Farooq et al. 2009). The quality and quantity of any plant growth depend on the aforementioned events, which can be severely affected by water deficit (Tardieu et al. 2018). A short-term water deficit affects the expansion rate, and this usually happens when crops are irrigated during the dry season (Heuer and Nadler 1995).

Water stress greatly restrains cell expansion and cell growth under low turgor pressure, which also affects the expansion of leaves. Water stress, which shrinks cells, causes a reduction in plant height (Jaleel et al. 2009). Moreover, water-limiting conditions result in impaired cell elongation, mainly because of the poor water flow from the xylem to the nearby cells (Nonami 1998). Reduced turgor pressure and the slow rate of photosynthesis under drought stress greatly limit leaf expansion (Rucker et al. 1995). The volume of leaves for any plant is influenced by water stress, and diminishing longevity and reduction of individual leaf size are affected by the reduction in soil water potential (Anjum et al. 2011). Moreover, water deficit has an adverse effect on crop production and plant growth which is caused by a reduction in fresh and dry biomass production (Zhao et al. 2006). Reduced leaf size is well correlated with drought stress, and indeed many xerophytes have developed small leaves during their adaptation to survive in severe environmental conditions. A small leaf area is advantageous to limit water use in plants and can be responsible for the low productivity of crops (Sinclair and Muchow 2001). They noted that different crops or genotypes behave differently.

Overall, all plants exposed to drought and suffering critical water deficit have significant morphological changes. For example, according to Mangena (2018), water deficit had a significant negative impact on the shoot and root morphology of soybean, including a reduction in the (i) number of new branches, (ii) initiation of leaves and expansion of the lamina, and (iii) number of trifoliate leaves. The reduction in shoot growth and root development caused a reduction in overall crop development and crop yield. Therefore, it is important to have robust agricultural management practices and drought mitigation strategies to minimize the impact of drought on crop morphology.

Effects of Drought on Crop Yields and Global Food Security

Challenges in ending hunger and food insecurity still exist, though extensive discussions have been ongoing to address the major causes of poverty and long-term hunger to reduce human anguish (Tanumihardjo et al. 2007; Haile 2005). The problem of drought onset has continued to receive close attention, given that it represents a key type of extreme climate event (Dai 2011), which causes loss of food production and, consequently, spikes in food prices (Lobell et al. 2011). The threats to global food security caused by climate change are one of the most critical challenges of the twenty-first century. While there is a need to supply adequate food for a growing global population, at the same time, there is also a need to sustain the already stressed environment. Availability of nutritious and quality food is an essential requirement for all humans, and agricultural sustainability is needed to ensure that the food demands of people are met (Brown and Funk 2008).

Although water stress may cause negative effects on overall growth and development of crops, the most significant impact of drought and water stress is a reduction of crop production, which contributes to the diminution of global food supplies (Lesk et al. 2016). Worldwide demand for food is anticipated to double by 2050 because of population growth, dietary change, and bioenergy use (Tilman et al. 2011), and an expected annual rate of yield increase of 2.4% will be necessary to meet this demand with existing farmlands (Fig. 3) (Ray et al. 2013). Meeting the growing need for food demand in the context of global warming requires better understandings of climate change and climatic factors, which influence crop production, and what is most important is to examine how crop yields respond to various climates and extremes. Adequately informed farmers are capable of adapting to the gradual changes in mean climate conditions, but for extreme events, there is a need for a better understanding of the impacts of climate extremes on crop production (Zampieri et al. 2017; Lesk et al. 2016). Drought, like an extreme weather event, will further harm crops and reduce yield (Lesk et al. 2016). Climate change has already caused critical effects on water resources such as irrigation and hydropower production (Beck and Bernauer 2011), food security, and human well-being. This is particularly noted in African countries but is currently beginning to involve the entire world (Magadza 2000).

Drought has aggravated the problem of food production because it is a global climatic threat that simultaneously influences food security (Haile 2005). Evaluating the impact of drought on crop production is difficult because drought itself is driven

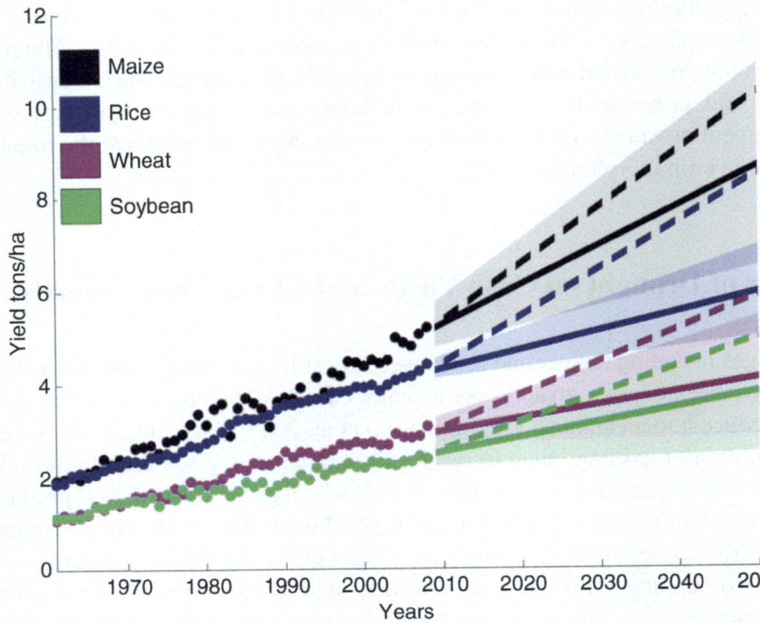

Fig. 3 Global projections: observed area-weighted global yield (1961–2008) shown using closed circles and projections to 2050 using solid lines for four crops (maize, rice, wheat, and soybean). Gray shading shows the 90% confidence region derived from 99 bootstrapped samples. The dashed line shows the trend of the ~2.4% yield improvement required each year to double production in these crops by 2050 without bringing additional land under cultivation starting in the base year of 2008. (Adapted from Ray et al. (2013)

by complex climatic conditions (Leng and Hall 2019). A crop failure during the rainy season is almost a complete agricultural failure, which reduces food availability at the household level as well as limits rural employment opportunities. If climate change acts to reduce crop production and, at the same time, populations increase, there is likely to be increasing hunger.

Agronomic Tools to Protect Crops from Drought

Agronomic tools used to mitigate the effects of drought on crops range from variety selection and the timing of seeding to cultural practices. Cultural practices include tillage and cultivation, crop production systems, mulching, fallowing, nutrient and irrigation management, and use of soil inoculants such as arbuscular mycorrhizal fungi (AMF) and plant growth-promoting rhizobacteria (PGPR) (Bodner et al. 2015; Creswell and Martin 1998; Parry et al. 2005). In addition, the exogenous application of protectants like glycine betaine and plant growth regulators has been useful for protecting crop plants under drought conditions (Farooq et al. 2009;

Lamaoui et al. 2018; Porcel et al. 2003; Porcel and Ruiz-Lozano 2004; Habibzadeh 2015).

Crop and Variety Selection Crop and variety selections most suited to the planting area are probably the most fundamental decisions to be made for crop production under drought conditions (Ferrante and Mariani 2018). Disregarding its importance has led to several crop failures in the past (Creswell and Martin 1998). Crop and variety selection for drought stress tolerance should be based on the tolerance level of the crop or variety, the time that the crop or variety takes to mature, and the characteristics which favor survival under drought conditions (Creswell and Martin 1998; Idowu et al. 2012). Early maturing crop varieties typically grow and mature before a drought reaches its peak during the growing season, while varieties with short stems with small leaf surface area can reduce transpiration. Similarly, varieties with deep and extensive root systems improve the capture and use of available soil moisture (Creswell and Martin 1998).

Time of Planting It is critical to choose the best time for seeding when cropping under dry conditions, because it helps match water availability to crop demand and optimizes crop establishment and early plant vigor (Bodner et al. 2015). Early sowing is encouraged in dry environments because it can improve the water use efficiency of crops (Brown et al. 1989; Eastham et al. 1999) and can ensure flowering and grain filling (both critical growth stages of crops) which occur during periods of better soil water availability (Herero and Johnson 1981). Early sowing also helps crops to develop deeper roots and avoid early droughts (Barraclough and Leigh 1984; Brown et al. 1989; Incerti and O'Leary 1990). Higher crop yields of wheat, barley, and rapeseed have also been attributed to early sowing in dry climates (Ehlers and Goss 2003; Kirkland and Johnson 2000; Latiri et al. 2010). On the other hand, late sowing could lead to reduced crop yield (Mahdi et al. 1998).

Stand Density Reducing stand density is another agronomic tool often explored for water saving in cropping systems situated in moisture-deficient environments (Bodner et al. 2015). Though this practice tends to (i) lower crop interception of solar radiation, (ii) increase evaporation losses of water and runoff, and (iii) increase weed competition, especially for crops with wide rows, it appears to be very effective at water savings and hence yield optimization under intermittent terminal stress levels (Bodner et al. 2015).

Tillage Practices Tillage practices impact on soil hydraulic properties, including soil hydraulic conductivity, implying that these practices can affect moisture storage in the soil. A review of literature on the influence of tillage on soil hydraulic properties (Bodner et al. 2015) revealed that reduced tillage tends to increase water storage in the soil through higher storage in fine pores in spite of reduced total porosity and macropore volume. They found that this trend applied to similar hydrological regimes and different soil textures. Bodner et al. (2015) reported that saturated hydraulic conductivity (for those which are macropore dependent) showed no unique trend in tillage experiments because the effects of tillage on soil macropores change over time. This suggests that knowledge of temporal variability is necessary for a full understanding of the effects of different tillage practices on soil moisture storage.

Crop Production Systems Polyculture or multiple crop production systems that control erosion, increase water and nutrient retention, and also have a potential to increase yield, should be employed for crop production under dry environments. Examples of these systems include crop rotation and strip cropping. Though crop rotation is typically more commonly practiced in humid regions, it can be useful in dry regions if crop rotations are planned around crop moisture requirements. In the Sahel regions of West Africa and dry regions of India, the inclusion of mulched fallows in crop rotations has significantly helped crop survival and hence healthy stand establishment (Creswell and Martin 1998). Crop rotations in these environments should also focus on selecting crops that help improve soil structure and the addition of organic matter to the soil to minimize soil erosion. These are typical in dry cropping environments (Bodner et al. 2015; Creswell and Martin 1998). Such planning can also maintain and/or improve the nutrient levels of soils in these environments. Strip cropping essentially involves planting crops in alternate strips which are usually planted perpendicular to slopes or the direction of prevailing winds to control erosion problems. Strip cropping also incorporates elements of crop rotation, contour cultivation, and stubble mulching, which are all good farming practices (Creswell and Martin 1998). Hence, the soil water storage potential of this approach is attributable to the combined benefits of all of these advantageous practices.

Fallowing Fallowing involves keeping the land free of vegetation for at least one growing season, with the intention of storing moisture gained from rainfall in the soil for use by a subsequent crop. In the US High Plains, alternating winter wheat with fallows has more than doubled wheat yields (Waldren 2003). Similarly, it is reported that maintaining about 2–2.4 ha of land each year, in summer fallows in India, has helped farmers to almost completely reduce drought-induced famine (Creswell and Martin 1998). It is noteworthy that for a fallow system to be successful, it must maintain high infiltration rates, protect the soil from erosion, and control weeds using good tillage practices that maintain sufficient residue on the soil surface (Waldren 2003; Creswell and Martin 1998). In this regard, the use of less stirring tillage practices, such as tine cultivation, the timing of tillage operations, and proper management of soil surface residues, are paramount.

Mulching and Stubble Tillage This technique involves covering up the soil surface with a protective layer, which may be organic or inorganic. Mulching helps hold moisture in the soil by reducing evaporation and runoff, which protects the soil and enhances its condition for supporting crop growth (Jabran 2019). High amounts of mulch (>50% of total straw produced by a crop field) are required for covering the soil surface, which is one of the demerits of mulches (Bodner et al. 2015; Kálmar et al. 2013). The extent to which mulching reduces evaporation is reported to range up to 28% (Zaongo et al. 1997; Eberbach et al. 2011), while moisture storage by mulched soils is documented to range between 8 and 22% (Kálmar et al. 2013; Jabran et al. 2015; Ramakrishna et al. 2006). Stubble tillage is also aimed at improving soil moisture storage and soil protection. However, it is more of a postharvest measure used during fallow periods between successive crops (Bodner et al. 2015). According to Creswell and Martin (1998), at least one ton of residue cover per hectare is required for stubble tillage to be effective. While they contend that this practice is beneficial with respect to water retention in the soil, other researchers (Kálmar et al. 2013; Unger et al. 1991) have reservations on its effectiveness as a water conserving management practice in semiarid areas.

Nutrient Management Studies have shown that proper nutrient management (at both macro and micro levels) can improve water use efficiency and promote crop yield (Farooq et al. 2017). Notable macronutrients include phosphorus and potassium, while important micronutrients include selenium, silicon, zinc, iron, and boron. Studies have shown that beans and sorghum grown during drought showed increased root growth, stomatal conductance, photosynthesis, membrane stability, and leaf water potential as a result of phosphorus nutrition (Alkaraki et al. 1996). Similarly, an adequate supply of potassium for grain legumes during drought conditions improved their tissue water potential and maintained photosynthesis at expected levels (Sangakkara et al. 2000). While selenium is reported to increase the ability of roots to uptake water under drought conditions (Farooq et al. 2014), silicon addition to drought-stressed plants increased their relative water content through increases in proline and glycine betaine (Hattori et al. 2005). Kurdali et al. (2013) have reported that the application of silicon alone or in combination with potassium to drought-stressed chickpea plants resulted in dry matter yield increases. The exogenous application of silicon has been reported to reduce the effects of drought in wheat and rice (Gong et al. 2005; Gautam et al. 2016). Besides increasing the relative water content of drought-stressed grains, applying zinc and iron can also positively affect their protein and micronutrient contents (Yadavi et al. 2014). Boron, on the other hand, is noted to improve the number and mass of nodules in soybeans grown under drought conditions when supplied through foliar application (Yamagishi and Yamamoto 1994).

Irrigation Since irrigation in cropping systems is not efficient and water wasted in the process is estimated to be over 50% of the amounts applied in some regions of the world (Parry et al. 2005), it is imperative that water use in crop production systems in dry environments is optimized. Water waste typically stems from technical issues associated with the distribution and inadequate maintenance of irrigation systems. This is often compounded by the high evapotranspiration and usually infertile fragile soils in dry environments that are prone to degradation and salinization (Parry et al. 2005; Ramoliya et al. 2004). Efficiency strategies include scheduling irrigation at night to reduce evapotranspiration, limiting overdependence on aquifers, and upgrading traditional irrigation systems to precision types coupled with precision agriculture (Parry et al. 2005). Other options include the use of recycled drainage water and gray water and irrigating crops during only critical growth stages as determined by crop requirements (Abu-Zeid and Hamdy 2002; Oweis et al. 1998; Araus et al. 2002; Parry et al. 2005). Another technique that has some documented success is partial root-zone irrigation or drying in which case irrigation is applied alternately to different sides of the root zone (Santos et al. 2003; de Souza et al. 2003; Loveys and Davies 2004).

Inoculating Soil with Arbuscular Mycorrhizal Fungi (AMF) and Plant Growth-Promoting Rhizobacteria (PGPR) Arbuscular mycorrhizal fungi (AMF) help plants resist drought through many mechanisms. First, they enhance water uptake from the soil through their extensive extra-radical mycelia (Porcel and Ruiz-Lozano 2004; Habibzadeh 2015). Second, AMF increases the antioxidant potential of plants under drought reducing lipid peroxidation in addition to producing more osmoprotectants (Porcel et al. 2003; Porcel and Ruiz-Lozano 2004; Habibzadeh 2015). The mechanisms by which plant growth-promoting rhizobacteria assist with plant drought stress resistance include solubilization of phosphorus, siderophore production, nitrogen fixation, and production of organic acids and plant growth enhancing substance and enzymes such as ACC deaminase, chitinase, and glucanase (Glick et al. 2007; Hayat et al. 2010). A listing of AMF and PGPRs that impact drought resistance in grain legumes is provided by Farooq et al. (2017).

Plant Growth Regulators (PGR) Plant growth regulators such as salicylic acid, cytokinins, and ABA are all reported to be involved in plant drought tolerance (Lamaoui et al. 2018). They help increase water potential and chlorophyll contents of plants under drought stress, which can all lead to crop yield increases (Zhang et al. 2004). In this regard, soybean yield increased when treated exogenously with ABA under drought conditions (Zhang et al. 2004). Transpiration is reported to have been reduced in potted miniature rose (*Rosa hybrida* L.) when applied with ABA in the spring or summer, and this was in addition to extended flower longevity (Monteiro et al. 2001). Foliar application of glycine betaine and salicylic on sunflowers improved their tolerance to drought. However, glycine betaine application was more effective s at the flowering stage (Hussain et al. 2008), suggesting a potential to increase sunflower yield under dry growing conditions.

Genetic Tools to Protect Crops from Drought

Drought is one of the most critical threats to crop production and agriculture in general. Under natural selection, various crop species have evolved to adapt to growth habitats of varying degrees of drought stress and are thus of different drought tolerance or water requirement. Information on general environmental requirements, including water requirements, and specific growth habit of a given crop can be easily obtained from the Ecocrop database,[1] which was established by the Food and Agriculture Organization (FAO) of the United Nations. Ever-continuing efforts on the breeding selection of improved varieties of all crop species for better yield, higher quality, and expanded cultivation environment since their domestication have overall been genetically enhancing their drought tolerance.

Drought tolerance is a complex multigene trait, and its genetic control and physiological mechanisms are yet to be fully understood. However, breeding for improvement of major crops, including wheat, maize, rice, and barley during the last century, has revealed many important characteristics of drought tolerance of these cereal crops responding to various selection practices. These lessons could serve as general guidance for future breeding efforts toward improvement of crop drought tolerance.

Some of these characteristics were illustrated in the generalized yield-versus-drought stress curves in Fig. 4. Of particular importance were the following observations:

1. Selections for yield increase under zero or moderate drought stress have also been successful in improvement of drought tolerance in new genotypes of higher yields (Araus et al. 2002; Slafer et al. 2005; Tambussi et al. 2005). This has been witnessed in rice and wheat (Serraj et al. 2011; Trethowan et al. 2002).
2. The selected higher-yield breed usually has an equal percentage improvement of drought tolerance under varying degrees of stress (Araus et al. 2002), exhibiting a larger yield increase in the absolute term under low drought stress conditions (Slafer et al. 1994).
3. For most crops, the selected higher-yield breed exhibited continued linear year-by-year genetic improvement of yield along with drought tolerance during a post-release multi-year cultivation period, as revealed by studies of grain yield increases in some barley and wheat genotypes commonly grown in the last century (Cattivelli et al. 2008; Slafer et al. 1994).
4. Direct selection for drought tolerance under moderate to severe drought stress has not been as successful in most crops due to polygenic control of the complex trait, epistasis, significant interactions between genotype and environment (G × E), and low heritability of selected traits (Piepho 2000). Therefore, drought tolerance of a crop may not be genetically enhanced without affecting the yield of reproductive organ of the crop. In other words, selection for genetic gains of yield in a crop without drought stress may be far better an approach for improvement of drought tolerance than those under drought stress. In addition, despite many emerging novel genetic and genomics approaches, the traditional breeding selection remains as a major genetic tool for new breeds with improved drought tolerance (Reviewed in Ashraf 2010).

Over the past half-century, research and crop improvement efforts in the area of drought tolerance have greatly furthered our understanding of physiological mecha-

[1] http://ecocrop.fao.org/ecocrop/srv/en/home

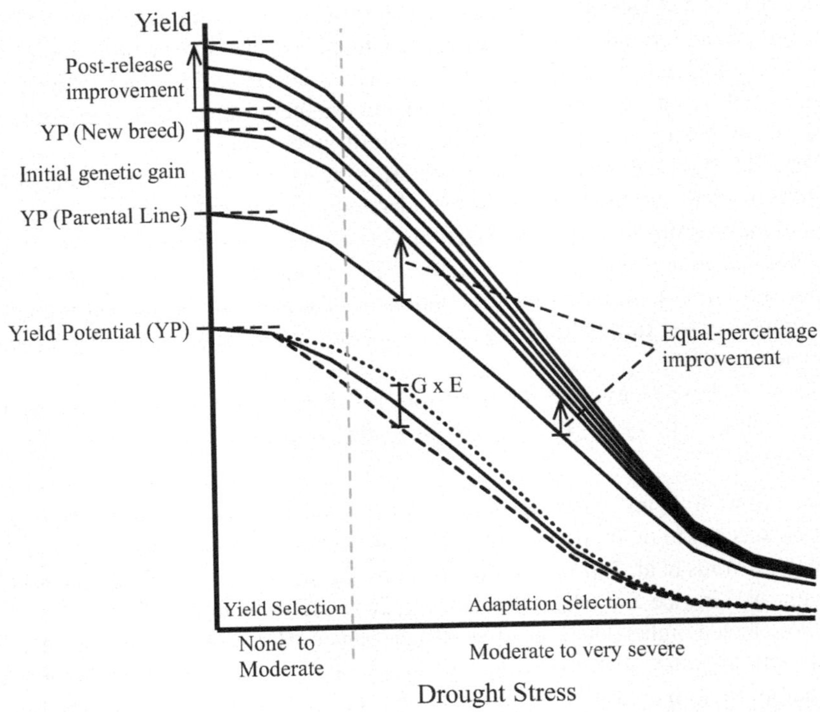

Fig. 4 Graphic illustration of the improvement of crop drought tolerance through breeding. The generalized bottom curves illustrate yield variations of a crop responding to the interaction of genotypes and environment (G × E) under varying degrees of drought stress. The top generalized curves illustrate yield improvement of a new breed grown in the initial release and post-release multiple years, as compared to that of a parental line, under varying degrees of drought stress. (Summarized from Reviews by Golldack et al. 2011; Xoconostle-Cazares et al. 2010; Chaves et al. 2003; Shinozaki and Yamaguchi-Shinozaki 2007)

nisms (Farooq et al. 2009; Golldack et al. 2011; Xoconostle-Cazares et al. 2010) and genetic control (Chaves et al. 2003; Shinozaki and Yamaguchi-Shinozaki 2007) of drought tolerance in various crops. These efforts have also led to the identification of lists of drought-responding physiological traits and tolerance-modulating genes in various crops (Cattivelli et al. 2008), and generated repertories of genetic resources, including genetic maps and transcriptome or genome sequences. Along with this advancement, two new approaches for crop improvement, (i) marker-assisted selection or breeding (MAS or MAB) and (ii) biotechnology involving direct genetic modification of target traits by transformation, have been developed and have been put to use. These two new approaches plus traditional breeding are currently the troika of genetic tools for drought tolerance improvement in all major crops and have generated long lists of new breeds of improved drought tolerance in various crops (Ashraf 2010).

Advancement of crop genomics over the recent decades has provided some new tools, especially molecular markers and genetic mapping methods for crop improvement. These genomics tools have facilitated a more efficient identification of desirable intraspecific genetic variations, including those at drought-stress-responding quantitative loci (QTLs) for drought tolerance. More effective transfers of these variations to generate new breeds of improved drought tolerance with the assistance of their associated molecular markers (i.e., MAB) have been also achieved in many crop species. A list of successful new breeds of improved drought tolerance in several crops was summarized by Ashraf (2010). All these new breeds exhibited increased yields, although the QTLs selected were mostly associated with drought tolerance improvement, mirroring the results from the breeding selection of yield improvement, as summarized in Fig. 4. However, this approach faces a major hurdle, which is the genetic constraint in a crop. Multivariate selections of multiple desirable QTLs in a new breed may not yield desired expression levels for all these quantitative traits, nor have the desired additive effect from the combination of these QTLs, due to the genetic constraint in the crop (Juenger 2013). A crop species that has evolved to adapt to an environment of a certain water availability range may thus be genetically constrained to a drought tolerance limit.

Theoretically, it is possible to enhance the drought tolerance limit of a crop by introducing foreign genetic materials by conferring added drought tolerance through genetic transformation, which is a biotechnological approach. There have been many genetically engineered crop lines with improved drought tolerance conferred by foreign genes expressing organic osmolytes, transcription factors, late embryogenesis proteins, and hormones (Juenger 2013). Nevertheless, it is unclear if the drought tolerance limit of these crops was actually enhanced, or if the improved drought tolerance was simply achieved by a new combination of intraspecific genetic variations through traditional breeding or MAB. Although the biotechnology approach for enhancing crop drought tolerance is a promising new technology (Deikman et al. 2012), it is currently not a cost-effective, nor publicly favorable, approach due to lengthy and costly research and development requirements, strict regulations, and unfavorable customer acceptance to genetically modified organisms.

It is worth mentioning that grafting, which is strictly a nongenetic tool, may be far more cost-effective and is thus still prevalent in the agricultural production of some vegetable and fruit crops. Grafting seedlings of vegetable crops such as some cucurbit species (cucumber, melon, and watermelon) and solanaceous crops (eggplant, pepper, and tomato) to rootstocks (e.g., special breeds of pumpkins), which have a stronger water-uptake capability, can (i) improve the drought tolerance of these crops, (ii) expand their cultivation to otherwise non-cultivable land, and (iii) enhance their tolerance to other abiotic stresses such as low temperature and resistance against some soil diseases such as root rot (Schwarz et al. 2010). The grafting approaches in these crops are not only cost-effective when compared to breeding and biotechnology approaches but are also currently irreplaceable in some crops for combatting certain root diseases, as no natural genetic variations conferring resistance against these diseases have been identified in these crop species (King et al. 2008).

Strategies for Drought Mitigation and Crop Management Under Changing Climate Conditions

Agricultural drought generally results from the deficiency of precipitation over an extended period of time that exacerbates dry conditions and leads to water stress, which causes a reduction in crop growth and development (Solh and van Ginkel 2014). Generally, drought is the result of a combination of below-average precipitation and above-average temperatures, which can be for a short duration (such as 1 week) or can persist across multiple years (McFadden et al. 2019). The potential effects of climate change on crop yield are on the increase, and it is necessary to make farming more resilient to climate extremes like drought. The impact of drought can be reduced through appropriate strategies (drought preparedness and mitigation strategies) and adapting the best agricultural management practices (crop rotation, growing drought-tolerant crops) under the changing climate scenario. Most farmers believe climate change is occurring, and they need to act on it because adaptation strategies at the farm level can contribute to counteracting these adverse climatic effects (Brumbelow and Georgakakos 2001). Building drought resilience to manage the impacts of climate change on human activities is the main responsibility of water managers, either in planning for weather extremities or optimizing long-term resource utilization (Muller 2007).

Drought Mitigation, Preparedness, and Adaptation

A drought mitigation plan is designed to reduce the impacts of drought by identifying the principal activities, groups, or regions most at risk (Wilhite et al. 2000). It is expected that climate change might increase or alter the intensity and frequency of droughts throughout the world in the future (Logar and van den Bergh 2013); thus, in the face of increasing uncertainties on the location, frequency, intensity, and duration of future drought, it is important to have a suite of better preparedness planning schemes, mitigation actions, and response strategies (Cai et al. 2015; Strzepek et al. 2010). It is widely accepted that drought impact can be minimized through preparedness and mitigation approaches. A better drought prediction system could help to mitigate the effects of drought, but although model performance has continued to improve, the general circulation models (GCM) used to predict climate change and associated drought parameters are mixed in their predictions for precipitation and temperature, which affect drought preparedness and mitigation (Cai et al. 2009).

According to Solh and van Ginkel (2014), drought cannot be prevented, but through better preparedness and mitigation actions, it is possible to minimize the impact of drought on crop production, develop more resilient ecosystems, and improve resilient systems to recover from the drought. Preparedness strategies are employed including geographical shift of agricultural systems (e.g., if a certain zone has high aridity, an appropriate cropping system can be adapted), climate-proofed rainfed cropping systems (growing drought-tolerant crops and their varieties), implementing high efficient irrigation system (improving efficiency of irrigation systems), and adapting combined rainfed and irrigated systems (Solh and van Ginkel 2014).

In addition, integrated approaches and strategies for better preparedness, mitigation, and adaptation are necessary to cope with future drought (Fig. 5). Moreover, drought policy should emphasize risk management through the implementation of best preparedness, mitigation, and adaptation (Wilhite 2002). Also, robust and effective monitoring systems, best management practices, and prediction and warning systems further help to reduce the impact of drought on crop production and development. In addition, efficient risk and impact assessment, response, and recovery systems will enhance the approaches to drought mitigation, preparedness, and adaptation strategies, not only during the drought period but also in acting to cope with future drought.

Drought-Resilient Agriculture

Although it is well recognized that drought is one of the major causes of crop yield reductions, limited options are available for farmers to minimize the damaging effects of drought (McFadden et al. 2019). Any mitigation actions that reduce drought risk and vulnerability will definitely increase resilience. For example, dur-

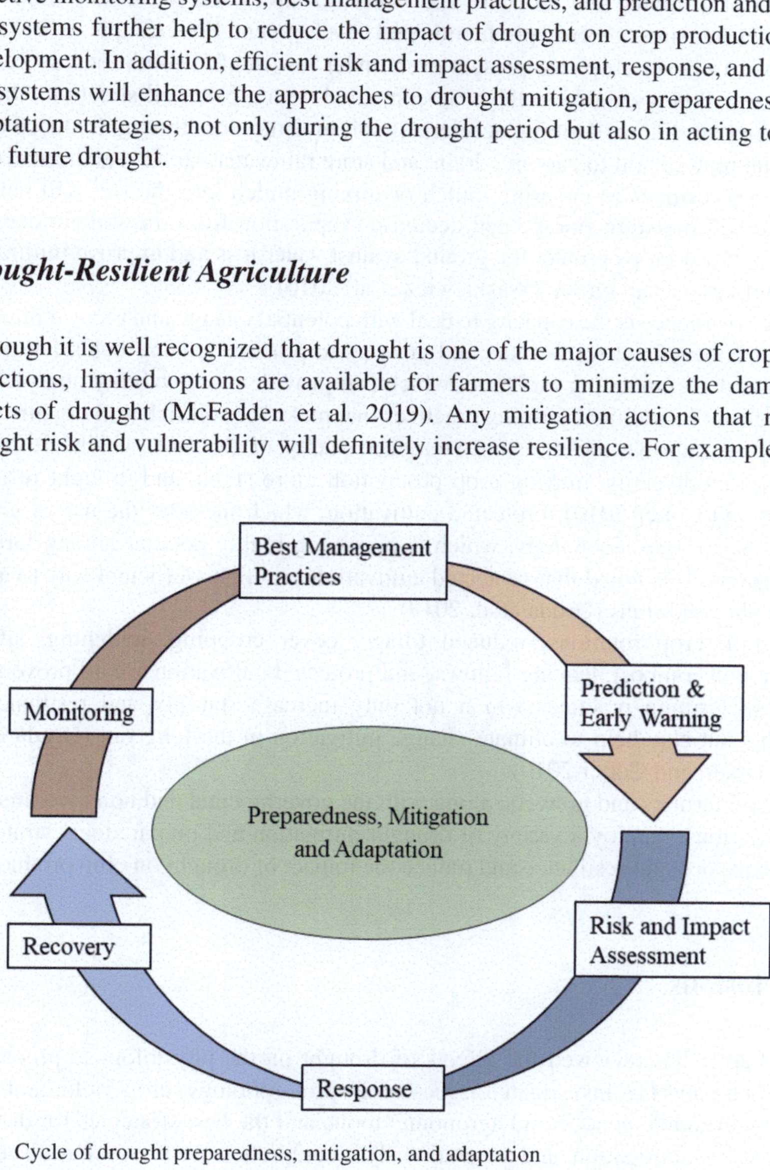

Fig. 5 Cycle of drought preparedness, mitigation, and adaptation

ing drought, it is essential at least to adapt to the best water conservation and crop management practices. For water conservation, farmers and growers have to use an efficient irrigation system, reduce water losses, and use nonconventional water resources for irrigation and plant crops with low water requirements. Similarly, for best crop management, farmers, growers, researchers, and governmental and non-governmental agencies need to work together to develop drought-tolerant crops and their varieties. They also need to reduce tillage and introduce crop rotation, mixed cropping, and cover cropping systems. These measures will lead to the better management of available soil moisture and water resources and reduce the impact of drought/water stress on crop production and development. For example, where residue cover or cover crops are present even under low rainfall conditions, more soil moisture will be available to the crop compared with a bare soil situation. On the other hand, traditional farming accelerates soil moisture loss through reduced ability of the plowed soil to capture, drain, and store rainwater. However, alternatively, using crop residues as covering mulch or mixing mulch into the soil will help to increase soil moisture storage and decrease evaporation from the soil surfaces. In addition, cover crops protect the ground against water loss and improve infiltration and limit water evaporation (Waskiewicz et al. 2016).

Since resilience is the capacity to deal with potential change and recover after the event, it is beneficial for farmers and growers to practice leaving fields fallow for resting and accumulating moisture, which can provide more stability and yield in the long run. In addition, farming practices that make the soil richer in organic matter help to improve the moisture storage capacity of the soil, which ultimately increases biodiversity, making crop production more stable and drought resilient (Tirado and Cotter 2010). Protected cultivation, which includes the use of greenhouses, is an agro-technology, which is becoming highly popular among farmers and growers. It is noted that protected cultivation is a highly efficient way to adapt to drought conditions (Gruda et al. 2019).

Overall, crop rotations, reduced tillage, cover cropping, mulching, adding manure and compost, leaving fallows, and protected cultivation are all proven and available farming practices which not only increase stability and resilience to droughts but also help to climate change mitigation in the long run (Gruda et al. 2019; Tirado and Cotter 2010).

Hence, farmers and growers, along with the governmental and nongovernmental agencies, must employ a variety of drought mitigation and preparedness strategies to enhance drought resilience and reduce the impact of drought on crop production.

Conclusions

This chapter has reviewed the effects of drought on the physiological process of crop plants and has investigated issues of crop morphology, crop yield, and food security, available genetic and agronomic tools, and the best strategies for drought preparedness, mitigation, and adaptation. This comprehensive review has discussed some of the critical issues that need to be addressed to protect crops under drought stress. This chapter implies that to reduce the impact of drought stress on crop development and production, best crop management practices, monitoring mechanisms, drought prediction and early warning systems, effective and timely risk and impact assessment, effective response, and recovery strategies, and appropriate genetic and agronomic tools, may need to be undertaken. In addition, knowledge of the relationship between climate change-induced agricultural drought and crop production will be critical for many decision-makers including farmers, growers, and governmental and nongovernmental agencies; therefore, it would be of utmost importance to implement educational and awareness programs for drought preparedness, mitigation, and adaptation strategies from a local to a global scale.

Climate change predictions suggest that there will be increased frequency and severity of such droughts, which gives an even greater sense of urgency to identify crops that are resilient and can produce under such adverse conditions (Motsa et al. 2015; Modi and Mabhaudhi 2013). It is also recommended to develop a plant hardiness zone map in each region, which helps to understand and select potential crops in a particular location for better management practice under changing climate. For example, information on intra-seasonal variability might be useful to adjust the crop planting season (Cai et al. 2009). To strengthen drought preparedness, mitigation, and adaptation strategies, governments and policymakers should increase their efforts to enhance research works to minimize the impact of climate extremes such as drought on agriculture. An integrated approach to the effects of drought on crop production, crop responses to drought, and potential strategies for drought preparedness, mitigation, and adaptation is necessary to help us better understand crop and drought management under drought stress.

Acknowledgments This work was supported by the Evans-Allen project of the US Department of Agriculture (USDA), National Institute of Food and Agriculture (NIFA).

References

Abu-Zeid M, Hamdy A (2002) Water vision for the twenty-first century in the Arab World. World Water Council. 3rd World Water Forum

Alkaraki GN, Clark RB, Sullivan CY (1996) Phosphorus nutrition and water stress effects on proline accumulation in sorghum and bean. J Plant Physiol 148:745–751

Anjum SA, Xie X-Y, Wang L-C, Saleem MF, Man C, Lei W (2011) Morphological, physiological and biochemical responses of plants to drought stress. Afr J Agric Res 6(9):2026–2032. https://doi.org/10.5897/AJAR10.027

Anjum SA, Ashraf U, Zohaib A, Tanveer M, Naem M, Ali I, Tabassum et al (2017) Growth and developmental responses of crop plants under drought stress: a review. Zemdirbyste-Agriculture 104(3):267–276. https://doi.org/10.13080/z-a.2017.104.034

Araus JL, Slafer GA, Reynolds MP, Royo C (2002) Plant breeding and drought in C3 cereals: what should we breed for? Ann Bot 89:925–940. https://doi.org/10.1093/aob/mcf049

Ashraf M (2010) Inducing drought tolerance in plants: recent advances. Biotechnol Adv 28:169–183. https://doi.org/10.1016/j.biotechadv.2009.11.005

Athar HR, Ashraf M (2005) Photosynthesis under drought stress. In: Pessarakli M (ed) Handbook of Photosynthesis. Taylor and Francis, Inc., New York, U.S.A., pp 793–809

Barraclough PB, Leigh RA (1984) The growth and activity of winter wheat roots in the field: the effect of sowing date and soil type on root growth of high-yielding crops. J Agric Sci 103:59–74

Beck L, Bernauer T (2011) How will combined changes in water demand and climate affect water availability in the Zambezi river basin? Global Environmental Change 21(3): 1061–1072. https://doi.org/10.1016/j.gloenvcha.2011.04.001

Bodner G, Nakhforoosh A, Kaul HP (2015) Management of crop water under drought: a review. Agron Sustain Dev 35:401–442

Brown ME, Funk CC (2008) Food security under climate change. Science 319(5863):580–581. https://doi.org/10.1126/science.1154102

Brown SC, Cooper PJ, Keatinge JD (1989) Root and shoot growth and water use of chickpeas (Cicer arietinum) grown in dryland conditions: effects of sowing date and genotype. J Agric Sci 113:41–49

Brumbelow K, Georgakakos A (2001) An assessment of irrigation needs and crop yield for the United States under potential climate changes. J Geophys Res 106(D2):27383–27405

Cai X, Wang D, Laurent R (2009) Impact of climate change on crop yield: a case study of rainfed corn in Central Illinois. J Meteorol Climatol 48:1868–1881

Cai X, Zeng R, Kang WH, Song J, Valocchi AJ (2015) J Water Resour Plan Manag 141(9):04015004

Cattivelli L, Rizza F, Badeck FW, Mazzucotelli E, Mastrangelo AM, Francia E, Stanca AM (2008) Drought tolerance improvement in crop plants: an integrated view from breeding to genomics. Field Crop Res 105(1–2):1–14. https://doi.org/10.1016/j.fcr.2007.07.004

Chaves MM, Maroco JP, Pereira JS (2003) Understanding plant responses to drought – from genes to the whole plant. Funct Plant Biol 30(3):239–264. https://doi.org/10.1071/Fp02076

Creswell R, Martin FW (1998) Dryland farming: crops and techniques for arid regions. Echo Technical Notes

Dai A (2011) Drought under global warming: a review. Wiley Interdiscip Rev Clim Chang 2:45–65. https://doi.org/10.1002/wcc.81

Deikman J, Petracek M, Heard JE (2012) Drought tolerance through biotechnology: improving translation from the laboratory to farmers' fields. Curr Opin Biotechnol 23(2):243–250. https://doi.org/10.1016/j.copbio.2011.11.003

Eastham JP, Gregory J, Williamson DR, Watson GD (1999) The influence of early sowing of wheat and lupin crops on evapotranspiration and evaporation from the soil surface in Mediterranean climate. Agric Water Manag 42:205–218

Eberbach PL, Humphreys E, Kukal SS (2011) The effect of rice straw mulch on evapotranspiration, transpiration and soil evaporation of irrigated wheat in Punjab, India. Agric Water Manag 98:1847–1855. https://doi.org/10.1016/j.agwat.2011.07.002.

Ehlers W, Goss M (2003) Water dynamics in plant production. CABI Publishing, New York

Fahad S, Bajwa AA, Nazir U, Anjum SA, Farooq A, Zohaib A, Sadia et al (2017) Crop production under drought and heat stress: plant responses and management options. Front Plant Sci 8:1147. https://doi.org/10.3389/fpls.2017.01147

Farooq M, Wahid A, Kobayashi N, Fujita D, Basra SM (2009) Plant drought stress: effects, mechanisms and management. Agron Sustain Dev 29:185–212. https://doi.org/10.1051/agro:2008021

Farooq M, Hussain M, Siddique KH (2014) Drought stress in wheat during flowering and grain-filling periods. Crit Rev Plant Sci 33:331–349

Farooq M, Gogoi N, Barthakur S, Baroowa B, Bharadwaj N, Alghamadi SS, Siddique KH (2017) Drought stress in grain legumes during reproduction and grain filling. J Agron Crop Sci 203:81–102

Fathi A, Tari DB (2016) Effect of drought stress and its mechanism in plants. Int J Life Sci 10(1):1–6. https://doi.org/10.3126/ijls.v10i1.14509

Ferrante A, Mariani L (2018) Agronomic management for enhancing plant tolerance to abiotic stresses: high and low values of temperature, light intensity and relative humidity. Horticulturae 4:21. https://doi.org/10.3390/horticulturae4030021

Gautam P, Lal B, Tripathi BR, Shahid M, Baig MJ, Raja R, Maharama et al (2016) Role of silica and nitrogen interaction in submergence tolerance of rice. Environ Exp Bot 125:98–109. https://doi.org/10.1016/j.envexpbot.2016.02.008
Glick BR, Cheng Z, Czarny J, Duan J (2007) Promotion of plant growth by ACC deaminase-producing soil bacteria. Eur J Plant Pathol 119:329–339
Golldack D, Luking I, Yang O (2011) Plant tolerance to drought and salinity: stress regulating transcription factors and their functional significance in the cellular transcriptional network. Plant Cell Rep 30(8):1383–1391. https://doi.org/10.1007/s00299-011-1068-0
Gong H, Zhu X, Chen K, Wang S, Zhang C (2005) Silicon alleviates oxidative damage of wheat plants in pots under drought. Plant Sci 169:313–321. https://doi.org/10.1016/j.plantsci.2005.02.023
Gruda N, Bisbis M, Tanny J (2019) Impacts of protected vegetable cultivation on climate change and adaptation strategies for cleaner production – a review. J Clean Prod 225:324–339
Habibzadeh Y (2015) Arbuscular mycorrhizal fungi in alleviation of drought stress on grain yield and yield components of mungbean (Vigna radiata L.) plants. Int J Sci 4:34–40
Haile M (2005) Weather patterns, food security and humanitarian response in sub-Saharan Africa. Philos Trans R Soc B 360:2169–2182. https://doi.org/10.1098/rstb.2005.1746
Harris D, Tripathi RS, Joshi A (2002) On-farm seed priming to improve crop establishment and yield in dry direct seeded rice. In: Pandey S, Mortimer M, Wade L, Tuong TP, Lopes K, Hardy B (Eds.), Direct seeding: Research Strategies and Opportunities. International Research Institute, Manila, Philippines, pp. 231–240
Hattori T, Inanaga S, Araki H, An P, Mortia S, Luxova M, Lux A (2005) Application of silicon enhanced drought tolerance in sorghum bicolor. Physiol Plant 123:459–466
Hayat R, Ali S, Amara U, Khalid R, Ahmed I (2010) Soil beneficial bacteria and their role in plant growth promotion: a review. Ann Microbiol 60:579–598
Herero MP, Johnson RR (1981) Drought stress and its effects on maize reproductive systems. Crop Sci 21:105–110
Heuer B, Nadler A (1995) Growth and development of potatoes under salinity and water deficit. Aust J Agric Res 46:10. https://doi.org/10.1071/AR9951477
Howden SM, Soussana JF, Tubiello FN, Chhetri N, Dunlop M, Meinke H (2007) Adapting agriculture to climate change. Proc Natl Acad Sci 104(50):19691–19696
Hund A, Ruta N, Liedgens M (2009) Rooting depth and water use efficiency of tropical maize inbred lines, differing in drought tolerance. Plant Soil 318:311–325. https://doi.org/10.1007/s11104-008-9843-6
Hussain M, Malic MA, Farooq M, Ashraft MY, Cheema MA (2008) Improving drought tolerance by exogeneous application of glycinebetaine and salicylic acid in sunflower. J Agron Crop Sci 194:193–199
Hussain HA, Hussain S, Khaliq A, Ashraf U, Anjum SA, Men S, Wang L (2018) Chilling and drought stresses in crop plants: implications, cross talk, and potential management opportunities. Front Plant Sci 9:393. https://doi.org/10.3389/fpls.2018.00393
Idowu J, Marsalis M, Flynn R (2012) Agronomic principles to help with farming during drought periods. Guide A-147. New Mexico State Extension Publication, Las Cruces
Incerti M, O'Leary GJ (1990) Rooting depth of wheat in the Victorian Mall. Aus J Exp Agric 30:817–824
IPCC (2019) Summary for policymakers. In: Shukla PR, Skea J, Calvo Buendia E, Masson-Delmotte V, Pörtner H-O, Roberts DC, Zhai P, Slade R, Connors S, van Diemen R, Ferrat M, Haughey E, Luz S, Neogi S, Pathak M, Petzold J, Portugal Pereira J, Vyas P, Huntley E, Kissick K, Belkacemi M, Malley J (eds) Climate change and land: an IPCC special report on climate change, desertification, land degradation, sustainable land management, food security, and greenhouse gas fluxes in terrestrial ecosystems. In press
Jabran K (2019) Role of mulching in pest management and agricultural sustainability. Springer, Cham. https://doi.org/10.1007/978-3-030-22,301-4

Jabran K, Ullah E, Hussain M, Farooq M, Zaman U, Yaseen M, Chauhan BS (2015) Mulching improves water productivity, yield and quality of fine rice under water-saving rice production systems. J Agron Crop Sci 201(5):389–400

Jaleel CA, Manivannan P, Sankar B, Kishorekumar A, Gopi R, Somasundaram R, Panneerselvam R (2007) Water deficit stress mitigation by calcium chloride in Catharanthus roseus: effects on oxidative stress, proline metabolism and indole alkaloid accumulation. Colloids Surf B Biointerfaces 60(1):110–116. https://doi.org/10.1016/j.colsurfb.2007.06.006

Jaleel CA, Manivannan P, Wahid A, Farooq M, Al-Juburi HJ, Somasundaram R, Panneerselvam R (2009) Drought stress in plants: a review on morphological characteristics and pigments composition. Int J Agric Biol 11(1):100–105. https://doi.org/08–305/IGC-DYT/2009/11–1–100–105

Juenger TE (2013) Natural variation and genetic constraints on drought tolerance. Curr Opin Plant Biol 16(3):274–281. https://doi.org/10.1016/j.pbi.2013.02.001

Kálmar T, Bottlik L, Kisić I, Gyuricza C, Birkás M (2013) Soil protecting effect of the surface cover in extreme summer periods. Plant Soil Environ 59:404–409

Kang Y, Khan S, Ma X (2009) Climate change impacts on crop yield, crop water productivity and food security – a review. Prog Nat Sci 19:1665–1674

Karl TR, Melillo M, Peterson TC (eds) (2009) Global climate change impacts in the United States. Cambridge University Press, New York

Kaya MD, Okçub G, Ataka M, Çıkılıc Y, Kolsarıcıa Ö (2006) Seed treatments to overcome salt and drought stress during germination in sunflower (*Helianthus annuus* L.). Eur J Agron 24:291–295

King SR, Davis AR, Liu WG, Levi A (2008) Grafting for disease resistance. HortScience 43(6):1673–1676

Kirkland KJ, Johnson EN (2000) Alternative seeding dates (fall and April) affected Brassica napus canola yield and quality. Can J Plant Sci 80:713–719

Kurdali F, Al-Chammaa M, Mouasess A (2013) Growth and nitrogen fixation in silicon and/or potassium fed chickpeas grown under drought and well-watered conditions. J Stress Physiol Biochem 9:385–406

Lamaoui M, Jemo M, Datia R, Bekkaoui F (2018) Heat and drought stresses in crops and approaches for their mitigation. Front Chem 6:26. https://doi.org/10.3389/fchem.2018.00026

Latiri K, Lhomme JP, Annabi M, Setter TL (2010) Wheat production in Tunisia: progress, inter-annual variability and relation to rainfall. Eur J Agron 33:33–42

Lehmann N (2013) How climate change impacts on local cropping systems: a bioeconomic simulation study for Western Switzerland. Ph.D. dissertation, Environmental Science, 169. https://doi.org/10.3929/ethz-a-009933856

Leng G, Hall J (2019) Crop yield sensitivity of global major agricultural countries to droughts and the projected changes in the future. Sci Total Environ 654:811–821

Lesk C, Rowhani P, Ramankutty N (2016) Influence of extreme weather disasters on global crop production. Nature 529:84–87. https://doi.org/10.1038/nature16467

Liang X-Z, Wu Y, Chambers RG, Schmoldt DL, Gao W, Liu C, Liu et al (2017) Determining climate effects on US total agricultural productivity. PNAS. https://doi.org/10.1073/pnas.1615922114

Lobell DB, Schlenker W, Costa-Roberts J (2011) Climate trends and global crop production since 1980. Science 333(6042):616–620. https://doi.org/10.1126/science.1204531

Logar I, van den Bergh JC (2013) Methods to assess costs of drought damages and policies for drought mitigation and adaptation: review and recommendations. Water Resour Manag 27:1707–1720

Loveys B, Davies WJ (2004) Physiological approaches to enhance water use efficiency in agriculture: exploiting plant signalling in novel irrigation practice. In: Bacon M (ed) Water use efficiency in plant biology. Blackwell Scientific Publications, Oxford, pp 113–141

Magadza CHD (2000) Climate change impacts and human settlements in Africa: prospects for adaptation. Environ Monit Assess 61:193–205

Mahdi L, Bell CJ, Ryan J (1998) Establishment and yield of wheat (Triticum turgidum L.) after early sowing at various depths in a semi-arid Mediterranean environment. Field Crop Res 58:187–196

Mangena P (2018) Water stress: morphological and anatomical changes in soybean (Glycine max L.) plants. In: Andjelkovic V (ed) Plant, abiotic stress and responses to climate change. IntechOpen, London. https://doi.org/10.5772/intechopen.72899

Manzoor H, Athar HUR, Rasul S, Kanwal T, Anjam MS, Qureshi MK, Bashir et al (2016) Avenues for improving drought tolerance in crops by ABA regulation: molecular and physiological basis. In: Ahmad P (ed) Water stress and crop plants: a sustainable approach. Wiley, Chichester

McFadden J, Smith D, Wechsler S, Wallander S (2019) Development, adoption, and management of drought-tolerant corn in the United States, EIB-204. U.S. Department of Agriculture, Economic Research Service

Modi AT, Mabhaudhi T (2013) Drought tolerance and water use of selected South African traditional and indigenous crops. WRC report no. 1771/1/13. Water Research Commission, Pretoria

Monteiro JA, Nell TA, Barrett JE (2001) Postproduction of potted miniature rose: flower respiration and single flower longevity. J Am Soc Hort Sci 126:134–139

Motsa NM, Modi AT, Mabhaudhi T (2015) Sweet potato (Ipomoea batatas L.) as a drought tolerant and food security crop. S Afr J Sci. 111(11/12), Art. #2014–0252, 8 pages. http://dx.doi.org/10.17159/sajs.2015/20140252

Muller M (2007) Adapting to climate change: water management for urban resilience. Environ Urban Int Inst Environ Dev 19(1):99–113

Nabi RB, Tayade R, Huddsin A, Kulkarni KP, Imran QMI, Mun B-G, Yun B-W (2019) Nitric oxide regulates plant responses to drought, salinity, and heavy metal stress. Environ Exp Bot 161:120–133

Nonami H (1998) Plant water relations and control of cell elongation at low water potentials. J Plant Res 111:373–382. https://doi.org/10.1007/BF02507801

Oweis T, Pala M, Ryan J (1998) Stabilizing rainfed wheat yields with supplemental irrigation and nitrogen in a Mediterranean climate. Agron J 90:672–681

Parry M, Flexas J, Medrano H (2005) Prospects for crop production under drought: research priorities and future direction. Ann Appl Biol 147:211–226

Paudel B, Acharya BS, Ghimire R, Dahal KR, Bista P (2014) Adapting agriculture to climate change and variability in chitwan: long-term trends and farmers' perceptions. Agric Res 3(2):165–174

Piepho HP (2000) A mixed-model approach to mapping quantitative trait loci in barley on the basis of multiple environment data. Genetics 156(4):2043–2050

Porcel R, Ruiz-Lozano JM (2004) Arbuscular mycorrhizal influence on leaf water potential, solute accumulation, and oxidative stress in soybean plants subjected to drought stress. J Exp Bot 55:1743–1750

Porcel R, Barea JM, Ruiz-Lozano JM (2003) Antioxidant activities in mycorrhizal soybean plants under drought stress and their possible relationship to the process of nodule senescence. New Phytol 157:135–143

Ramakrishna A, Tam HM, Wani SP, Long TD (2006) Effect of mulch on soil temperature, moisture, weed infestation and yield of groundnut in Northern Vietnam. Field Crop Res 95:115–125. https://doi.org/10.1016/j.fcr.2005.01.030.

Ramoliya PJ, Patel HM, Pandey AN (2004) Effect of salinisation of soil on growth and macro- and micronutrient accumulation in seedlings of Acacia catechu (Mimosaceae). Ann Appl Biol 144:321–331

Ray DK, Mueller ND, West PC, Foley JA (2013) Yield trends are insufficient to double global crop production by 2050. PLoS One 8:e66428. https://doi.org/10.1371/journal.pone.0066428

Ray RL, Fares A, Risch E (2018) Effects of drought on crop production and cropping areas in Texas. Agric Environ Lett 3:170037

Razmjoo K, Heydarizadeh P, Sabzalian MR (2008) Effect of salinity and drought stresses on growth parameters and essential oil content of *Matricaria chamomile*. Int J Agric Biol 10:451–454

Rejeb KB, Benzarti M, Debez A, Savoure A, Abdelly C (2016) Water stress in plants: from gene to biotechnology. In: Ahmad P (ed) Water stress and crop plants: a sustainable approach. Wiley, Chichester

Riaz A, Younis A, Taj AR, Karim A, Tariq U, Munir S, Riaz S (2013) Effect of drought stress on growth and flowering of marigold (Tagetes erecta L.). Pak J Bot 45(S1):123–131

Rucker KS, Kvien CK, Holbrook CC, Hook JE (1995) Identification of peanut genotypes with improved drought avoidance traits. Peanut Sci 22(1):14–18. https://doi.org/10.3146/pnut.22.1.0003

Sangakkara UR, Frehner M, Nosberger J (2000) Effect of soil moisture and potassium fertilizer on shoot water potential, photosynthesis and partitioning of carbon in mungbean and cowpea. J Agron Crop Sci 185:201–207

Santos TP, Lopes CM, Rodrigues ML, Souza CR, Maroco JP, Pereira JS, Silva et al (2003) Partial rootzone drying: effects on growth and fruit quality of field-grown grapevines (Vitis vinifera). Funct Plant Biol 30:663–671

Sarker BC, Hara M, Uemura M (2005) Proline synthesis, physiological responses and biomass yield of eggplants during and after repetitive soil moisture stress. Sci Hortic 103:387–402

Schwarz D, Rouphael Y, Colla G, Venema JH (2010) Grafting as a tool to improve tolerance of vegetables to abiotic stresses: thermal stress, water stress and organic pollutants. Sci Hortic 127(2):162–171. https://doi.org/10.1016/j.scienta.2010.09.016

Serraj R, McNally KL, Slamet-Loedin I, Kohli A, Haefele SM, Atlin G, Kumar A (2011) Drought resistance improvement in rice: an integrated genetic and resource management strategy. Plant Prod Sci 14(1):1–14. https://doi.org/10.1626/pps.14.1

Shao H, Liye C, Mingan S, Jaleel, CA, Hongmei (2008) Higher plant antioxidants and redox signaling under environmental stresses. Comptes Rendus Biologies 331(6): 433–441. https://doi.org/10.1016/j.crvi.2008.03.011

Shinozaki K, Yamaguchi-Shinozaki K (2007) Gene networks involved in drought stress response and tolerance. J Exp Bot 58(2):221–227. https://doi.org/10.1093/jxb/erl164

Silva EC, Nogueira RJ, Vale FH, Melo NF, Araujo FP (2009) Water relations and organic solutes production in four umbu tree (Spondias tuberosa) genotypes under intermittent drought. Braz J Plant Physiol 21(1):43–53. https://doi.org/10.1590/S1677-04202007000300003

Silva EC, Albuquerque MB, Azevedo Neto AD, Junior CD (2013) Drought and its consequences to plants – from individual to ecosystem. In: Akıncı S (ed) Responses of organisms to water stress. IntechOpen. https://doi.org/10.5772/53833

Sinclair TR, Muchow RC (2001) System analysis of plant traits to increase grain yield on limited water supplies. Agron J 93:263–270. https://doi.org/10.2134/agronj2001.932263x

Sircelj H, Tausz M, Grill D, Batic F (2005) Biochemical responses in leaves of two apple tree cultivars subjected to progressing drought. J Plant Physiol 162:1308–1318

Slafer GA, Satorre EH, Andrade FH (1994) Increases in grain yield in bread wheat from breeding and associated physiological changes. In: Slafer GA (ed) Genetic improvement of field crops. Marcel Dekker Inc., New York, pp 1–68

Slafer GA, Araus JL, Royo C, Del Moral LFG (2005) Promising eco-physiological traits for genetic improvement of cereal yields in Mediterranean environments. Ann Appl Biol 146(1):61–70. https://doi.org/10.1111/j.17447348.2005.04048.x

Solh M, van Ginkel M (2014) Drought preparedness and drought mitigation in the developing world's drylands. Weather Clim Extrem 3:62–66

Souza CR, Maroco JP, Santos TP, Rodrigues ML, Lopes CM, Pereira JS, Chaves MM (2003) Partial rootzone drying: regulation of stomatal aperture and carbon assimilation in field-grown grapevines (Vitis vinifera cv. Moscatel). Funct Plant Biol 30:653–662

Specht JE, Chase K, Macrander M, Graef GL, Chung J, Markwell JP, Germann et al (2001) Soybean response to water. A QTL analysis of drought tolerance. Crop Sci 41:493–509. https://doi.org/10.2135/cropsci2001.412493x

Strzepek K, Yohe G, Neumann J, Boehlert B (2010) Characterizing changes in drought risk for the United States from climate change. Environ Res Lett 5

Tambussi EA, Nogues S, Ferrio P, Voltas J, Araus JL (2005) Does higher yield potential improve barley performance in Mediterranean conditions? A case study. Field Crop Res 91(2–3):149–160. https://doi.org/10.1016/j.fcr.2004.06.002

Tandzi LN, Bradley G, Mutengwa C (2019) Morphological responses of maize to drought, heat and combined stresses at seedling stage. J Biol Sci 19(1):7–16. https://doi.org/10.3923/jbs.2019.7.16
Tangu NA (2014) Effects on plant morphology of drought in olive. Turk J Agric Nat Sci 1:900–904
Tanumihardjo SA, Anderson C, Kaufer-Horwitz M, Bode L, Emenaker NJ, Haqq AM, Satia JA et al (2007) Poverty, Obesity, and Malnutrition: An International Perspective Recognizing the Paradox. J Am Diet Assoc. 107:1966–1972. https://doi.org/10.1016/j.jada.2007.08.007
Tardieu F, Simonneau T, Muller B (2018) The physiological basis of drought tolerance in crop plants: a scenario-dependent probabilistic approach. Annu Rev Plant Biol 69:733–759. https://doi.org/10.1146/annurev-arplant-042817-040218
Tátrai ZA, Sanoubar R, Pluhár Z, Mancarella S, Orsini F, Gianquinto G (2016) Morphological and physiological plant responses to drought stress in Thymus citriodorus. Int J Agron:1–8. https://doi.org/10.1155/2016/4165750
Tilman D, Balzer C, Hill J, Befort BL (2011) Global food demand and the sustainable intensification of agriculture. Proceedings of the National Academy of Sciences of the USA (PNAS) 108(50): 20260–20264. https://doi.org/10.1073/pnas.1116437108
Tirado, R.; Cotter, J. Ecological farming: drought-resistant agriculture. Greenpeace Research Laboratories, Greenpeace International, Amsterdam 2010.
Trethowan RM, van Ginkel M, Rajaram S (2002) Progress in breeding wheat for yield and adaptation in global drought affected environments. Crop Sci 42(5):1441–1446. https://doi.org/10.2135/cropsci2002.1441
Troy TJ, Kipgen C, Pal I (2015) The impacts of climate extremes and irrigation on US crop yields. Environ Res Lett 10:054013
Unger P, Stewart BA, Parr JF, Singh RP (1991) Crop residue management and tillage methods for conserving soil and water in semi-arid regions. Soil Tillage Res 20:219–240. https://doi.org/10.1016/0167-1987(91)90041-U.
Vico G, Dralle D, Feng X, Thompson S, Manzoni S (2017) How competitive is drought deciduousness in tropical forests? A combine eco-hydrological and eco-evolutionary approach. Environ Res Lett 12:065006
Waldren RP (2003) Introductory crop science. Pearson Custom Publishing, Boston
Waskiewicz A, Gladysz O, Beszterda M, Golinski P (2016) Water stress and vegetable crops. In: Ahmad P (ed) Water stress and crop plants: a sustainable approach. Wiley, Chichester
Wilhite DA (2002) Combating drought through preparedness. Nat Res Forum 26:275–285
Wilhite DA, Hayes MJ, Knutson C, Smith KH (2000) Planning for drought: moving from crisis to risk management. J Am Water Resour Assoc 36(4):697–710
Wu QS, Xia RX, Zou YN (2008) Improved soil structure and citrus growth after inoculation with three arbuscular mycorrhizal fungi under drought stress. Eur J Soil Biol 44:122–128. https://doi.org/10.1016/j.ejsobi.2007.10.001
Xoconostle-Cazares B, Ramirez-Ortega FA, Flores-Elenes L, Ruiz-Medrano R (2010) Drought tolerance in crop plants. Am J Plant Physiol 5(5):241–256
Yadavi A, Aboueshaghi RS, Dehnavi MM, Balouchi H (2014) Effect of micronutrients foliar application on grain qualitative characteristics and some physiological traits of bean (Phaseolus vulgaris L.) under drought stress. Ind J Fund Appl Life Sci 4:124–131
Yamagishi M, Yamamoto Y (1994) Effects of boron on nodule development and symbiotic nitrogen fixation in soybean plants. Soil Sci Plant Nutr 40:265–274
Yokota A, Kawasaki S, Iwano M, Nakamura C, Miyake C, Akashi K (2002) Citrulline and DRIP-1 protein (ArgE homologue) in drought tolerance of wild watermelon. Ann Bot 89:825–832
Yordanov I, Velikova V, Tsonev T (2000) Plant responses to drought, acclimation, and stress tolerance. Photosynthetica 38(1):171–186
Yordanov I, Velikova V, Tsonev T (2003) Plant responses to drought and stress tolerance. Bulg J Plant Physiol:187–206

Zampieri M, Ceglar A, Dentener F, Toreti A (2017) Wheat yield loss attributable to heat waves, drought and water excess at the global, national and subnational scales. Environ Res Lett 12:064008. https://doi.org/10.1088/1748-9326/aa723b

Zaongo CG, Wendt CW, Lascano RJ, Juo AS (1997) Interactions of water, mulch and nitrogen on sorghum in Niger. Plant Soil 197:119–126. https://doi.org/10.1023/A:1004244109990.

Zhang M, Duan L, Zhai Z, Li J, Tian X, Wang B (2004) Effects of plant growth regulators on water deficit-induced yield loss in soybean. In: Proceedings of the fourth international Crop Science Congress, Brisbane, QLD

Zhao T-J, Sun S, Liu Y, Liu J-M, Liu Q, Yan Y-B, Zhou H-M (2006) Regulating the Drought-responsive Element (DRE)-mediated signaling pathway by synergic functions of trans-active and trans-inactive DRE binding factors in Brassica napus. J Biol Chem 281(16):10,752–10,759. https://doi.org/10.1074/jbc.M510535200

Impact of Climate Change on Crop Production: Effects and Management

Azeem Tariq and Muhammad Adil Rashid

Introduction

One of the key challenges to agriculture in the twenty-first century is to feed the rapidly growing world population while adapting to the already deteriorating global climate (Lal 2005). Climate change has already affected crop productivity, water resources and food security globally but more severely in the developing world (Magadza 2000). Climate variabilities, for instance, high temperature, rise in carbon dioxide (CO_2) concentration in the atmosphere and changes in precipitation patterns, are causing year to year variation in crop growth and productivity, even in the areas with high-tech agricultural facilities (Reddy and Pachepsky 2000). The virtual conviction that earth's climate will continue to change raises many issues concerning agricultural productivity and crop quality. Hence, it is highly important and relevant to determine the influences of climate change on productivity and quality of crop plants in order to determine the feasible strategies to adapt to changing climate.

Climate change has fairly been rapid during the past few decades in many agricultural regions around the world. Climate change is affecting the crop productivity and expected to continue, if proper adaptation and mitigation strategies are not taken. There are four main climatic factors that represent climate change: rising atmospheric temperature, changes in precipitation patterns, elevated CO_2 concentration and increasing tropospheric ozone (O_3) level (Lobell and Gourdji 2012).

The authors Azeem Tariq and Muhammad Adil Rashid have been equally contributed to this chapter.

A. Tariq (✉) · M. A. Rashid (✉)
Department of Plant and Environmental Sciences, University of Copenhagen,
Copenhagen, Denmark
e-mail: azeem@plen.ku.dk; mar@plen.ku.dk

K. Jabran et al. (eds.), *Crop Protection Under Changing Climate*,
https://doi.org/10.1007/978-3-030-46111-9_7

The world temperature has been rising, with an average trend (rise) of 0.3 °C and 0.2 °C per decades for maximum and minimum temperature, respectively. The increasing temperatures are linked with high frequency of hot events and a low frequency of cold events, which influence crop growth and quality (Alexander et al. 2006). It is expected that global temperature could rise at the rate of 1 °C over the next 10-year period, which interprets as much as 2 °C increase in temperature in major agricultural regions, since lands heat up faster than oceans (Easterling and Wehner 2009). Temperature affects the crop growth and yield in many ways including its effects on photosynthesis, water use, crop duration, direct damage to plant tissues, and indirectly by increasing pest attacks (Ziska et al. 2011; Lobell and Gourdji 2012).

Changes in precipitation have direct consequences for crop production. Soil moisture as of direct relevance for crop growth is greatly influenced by variations in atmospheric temperature and seasonal precipitation. Generally, a substantial increase in drought and water scarcity has been projected for South and East Asia, Africa, Southern Europe and Eastern Australia (Sheffield and Wood 2008; Dai 2011). Increased drought events will lead to rise in agricultural crop water stress. Generally, plants respond to drought by closing their stomata, which reduce the net carbon uptake and increase the heat-related impacts, such as lower crop growth and quality. Water stress during the reproductive growth stage of crop plants severely affects the grain yield and quality. Alternatively, more intense rain events lead to flood and waterlogging, which destroy the crops and fertility of agricultural soils (Hatfield et al. 2011).

The atmospheric concentrations of CO_2 have been rising since industrial revolution, with an average rise of 2 ppm per annum in the 2000s (Peters et al. 2012). In 2018 atmospheric concentration of CO_2 (407.4 ppm) was 46.5% higher than at the begining of the industrial revolution (278 ppm). Atmospheric CO_2 concentrations are expected continue to increase in the next century at the rate of 25 ppm per decade, which will lead to a CO_2 level of 500 ppm by the mid of twenty-first century (IPCC 2001). The global rise in CO_2 concentration gives counteracting trends to the otherwise adverse impacts of rising temperature and increasing drought events. The expected increase in CO_2 concentrations will increase the global crop yield roughly by 1.8% per decade. At the same time, a rise in temperature is expected to decrease the crop yield roughly by 1.5% per decade. Elevated CO_2 may directly affect crop growth, development and quality, however, with different rates and magnitudes for different plant types (e.g. C_3, C_4) (Ainsworth and Long 2005; Taub and Wang 2008).

Air pollutants (e.g. nitrogen oxides, carbon monoxides, methane) form tropospheric ozone (O_3) by reacting with hydroxyl radicals in sunlight. The tropospheric O_3 concentrations have also increased from 10–15 ppm in the preindustrial age to over 35 ppm because of high emissions of pollutants from industrial activities. The predictions of the future O_3 level are uncertain due to improbability associated with ways of O_3 emission and pollutant control measures (Cape 2008).

In recent decades, significant studies have focused on enhancing our understanding of the impacts of climatic variables on crop production and mechanisms by which plants respond to these conditions (Kjøhl et al. 2011; Sangeetha et al. 2018).

The effects of climatic factors are complex, especially if multiple factors are combined. The interaction between climatic factors creates immense complexity that leads to either under or overestimation of the effects of an individual variable. In order to adapt to current and future climatic conditions, crop production strategies need to be modified. This chapter summarizes the most commonly known effects of changing climate on crop productivity and quality and management practices and technologies needed for adaptation to climate change.

Climate Change Effects on Crop Productivity and Quality

The primary climate variables that affect crop production and are associated with climate change are temperature, CO_2 and precipitation. Here, we briefly describe and discuss different ways by which these three major climatic variables affect crop productivity and quality.

Temperature

The increased climatic variability with a higher probability of occurrence of harmful weather events will make the conditions less suitable for crop production (Field et al. 2012). Climate change-induced changes in temperature dynamics can be multifaceted, that is, increased seasonal mean temperatures for certain crops and high-temperature episodes (heat waves), and to a lesser extent cold waves. Crops and crop growth stages that are sensitive to high temperature will become more vulnerable under such conditions. The main effects associated with the high temperature include leaf senescence, reduction in grain filling duration, perturbed leaf water relations and photosynthetic inhibition (Farooq et al. 2014). High temperature hastens crop development, ultimately shortening the crop duration. Higher than the optimum temperature during the post-anthesis period is particularly lethal for cereal crops where the grain-filling period is shortened, often resulting in lower yields (Asseng et al. 2015). Temperature also affects crop productivity by influencing the rate of photosynthesis and respiration. Photosynthesis is a complex process involving many enzymes and steps that have different temperature sensitivities. As such, the main enzyme involved in photosynthesis (rubisco) is very stable even at extreme high temperatures (Salvucci et al. 2001); however, rubisco activity declines under heat stress, because 'rubisco activase' (enzyme involved in activation of rubisco) is heat-labile (Salvucci et al. 2001; Salvucci and Crafts-Brandner 2004). Temperature effects on photosynthesis have also been observed in crops with a C_4 photosynthetic pathway (Crafts-Brandner and Salvucci 2000). Higher temperature also increases rates of respiration, which ultimately lowers the net carbon assimilation rate. The elevated temperature during the anthesis stage can cause pollen death and sterility, with a significant reduction in yields (Chavan et al. 2019).

Temperature-induced rise in vapour pressure deficit (VPD) is another aspect, which is often ignored in most studies; ultimately, it could lead to an exaggeration of temperature effects. It is known that as the temperature increases and absolute humidity remains constant, the leaf to air VPD increases exponentially. Increased VPD can have different consequences depending on crop response and microclimate. High VPD increases atmospheric deficit leading to an increase in evapotranspiration (ET). Moreover, the conditions in which soil water is limited, high ET may cause edaphic drought, thus, amplifying the overall effects of high temperature. Studies have shown that even a few degree increase in temperature can lead to plant death solely due to increased ET and water shortage (Will et al. 2013). Lobell et al. (2013) also indicated that temperature-induced rise in VPD acts as dual stress, because it not only increases water demand but also affects the water supply under rainfed agriculture, or at places where crop production depends on stored soil moisture. In a modelling study, they found that the high-temperature effects on maize through heat stress were non-existent and yield reduction was attributed to VPD-driven changes in water demand and supply. The VPD effects on ET are not straightforward and difficult to generalize. This is because plants tend to bring physiological changes in response to external stimuli, i.e. drought and VPD (Munns et al. 2010). It is known that high temperature reduces net photosynthesis (Law and Crafts-Brandner 1999; Monneveux et al. 2003); however, the photosynthetic inhibition could also be attributed to VPD-induced drought or stomatal closure and not just to direct heat effects. Further, increased VPD also lowers water use efficiency because more water is lost per molecule of CO_2 fixed.

Warming may also have a positive effect on crop yields in areas where crop production is restricted due to frost periods, and low temperature is a limiting factor to attain higher biomass and yields such as in North America (Izaurralde et al. 2003), Sweden (Eckersten et al. 2001), Northern Canada (Lesk et al. 2016) and China (Rashid et al. 2019). Higher seasonal mean temperature and reduction in the frost-free period may extend the cropping seasons in these areas. For example, low spring temperature slows the establishment of winter wheat in the north China plain, which can benefit from warming with fast establishment after overwintering (Rashid et al. 2019). Warming is also expected to help shifting crops towards those areas that are not suitable for crop production under current climatic conditions. For instance, warming would make the conditions suitable for maize production for grain purposes in Scandinavian countries.

Carbon Dioxide

Increasing atmospheric CO_2 concentration is a concern because it is one of the potent greenhouse gases responsible for global warming. Elevated CO_2 is likely to have significant direct effects on crop production through changes in physiology, growth, yield and even chemistry of plants. The magnitude and direction of these effects may also depend on other factors such as crop type and environmental conditions. Elevated CO_2 is expected to have fertilization-like effects. Net carbon assimilation rates are expected to increase with higher CO_2, through a reduction in photorespiration in C_3 crops (Ainsworth and Long 2005). Experimental evidence from free-air CO_2 enrichment (FACE) studies indicated that elevated CO_2 (500–600 ppm) for a number of crop species increases photosynthetic rate around 40% (Ainsworth and Rogers 2007).

The positive effect of elevated CO_2 through increased photosynthetic rate is much more pronounced in plants having C_3 than C_4 photosynthetic pathway. FACE and climate chamber studies have indicated that, in general, C_4 plants are less responsive to elevated CO_2 because the CO_2 concentration at the photosynthetic site is not a limiting factor even under the current ambient CO_2 level. The effects of elevated CO_2 on photosynthesis, biomass (Ainsworth and Long 2005) and yield of C_4 crops (Long et al. 2006) are much lower than for C_3 crops.

Legume crops are expected to benefit most out of elevated CO_2 (Rogers et al. 2009). Since, the positive effect of elevated CO_2 also depends on the availability and uptake of nitrogen, the effect is greater if nitrogen is not a limiting factor. Legumes have a specialized way to 'fix' atmospheric nitrogen through bacteria that live in nodules attached to plant roots. Therefore, legumes have been reported to have a greater increase in photosynthesis (Rogers et al. 2009) and less reduction in tissue nitrogen content at elevated CO_2 compared to nonleguminous crops (Taub and Wang 2008). Likewise, compared to rice, the photosynthesis and growth of soybean were significantly higher under elevated CO_2 in a FACE experiment (Long et al. 2006).

Another major effect of elevated CO_2 is regulation of plant stomata. Stomata are pore-like openings on plant surfaces, through which plants exchange gases (water vapour and CO_2) with the external environment. Among the stimuli affecting stomatal regulation, CO_2 is very crucial. Since stomata are the only gateway for CO_2 influx and water efflux, stomatal regulation to maintain CO_2 diffusion into leaves for photosynthesis and at the same time reducing water loss (transpiration) is a very delicate business. Elevated CO_2 has been shown to reduce stomatal conductance (a measure of stomatal openness) by an average of 22% across FACE experiments (Ainsworth and Rogers 2007). With the increase in CO_2 concentration, plants are expected to maintain higher photosynthetic rates even at low stomatal conductance. In general, lower stomatal conductance due to elevated CO_2 is expected to increase crop water use efficiency. Unlike the effects of elevated CO_2 on net photosynthesis, which are more pronounced in C_3 crops, its effects on stomatal conductance and water use efficiency are equally important for both C_3 and C_4 crops, and plant water use has been reported to decrease up to 20% (Leakey et al. 2009). Although, elevated CO_2 induced reduction in stomatal conductance has the potential to reduce water loss and in turn increase water use efficiency, the positive effect of elevated CO_2 on crop growth, e.g. leaf area and canopy size, could offset this effect. Higher leaf area and bigger canopy size could offer larger surface area for transpiration, thereby neutralizing the effect through reduced conductance. Lower stomatal conductance and transpiration can also affect leaf and canopy temperature. Cooler canopies are vital for saving plants from heat stress at high temperatures, thus, at elevated CO_2; reduced transpiration can jeopardize this heat tolerance mechanism. Moreover, lush green larger canopies and enhanced shading provide a suitable environment for pest proliferation, which could be another indirect effect under elevated CO_2.

The above-mentioned effects of elevated CO_2 on stomatal conductance and photosynthesis are clear; however, how much of these effects can actually translate into harvestable yield benefits is the main question. Literature indicates that these effects are highly variable, depending on other environmental conditions and crop types (Long et al. 2006). With higher photosynthetic rates and higher availability of photosynthates, plants tend to grow faster at elevated CO_2. The increased growth rate has been shown to increase average dry matter of plant by 17% for aboveground and 30% for belowground parts (De Graaff et al. 2006). Likewise, an increase in crop yield has also been reported at elevated CO_2 in FACE studies, where rice, wheat and soybean are the main candidates (Ainsworth 2008).

Crop Quality Under Elevated CO_2 and Temperature

Elevated CO_2 is also expected to affect the quality and chemical composition of plant organs. The main effects are the increase in leaf sugars and starch concentration and a decrease in leaf nitrogen concentration. FACE studies have indicated an increase in starch/sugars on average by 30–40% (Ainsworth 2008) and a reduction of leaf nitrogen concentration (per unit leaf mass) by 13% at elevated CO_2 (Ainsworth and Long 2005). Since tissue nitrogen status is closely related to protein concentration in plant organs, elevated CO_2 is likely to affect the nutritional quality and value of crops. The grain protein concentration of barley, rice and wheat has been reported to decrease at elevated CO_2 (Taub et al. 2008). Other than protein, elevated CO_2 has also reportedly a negative effect on the concentration of many other important minerals in plants such as phosphorus, magnesium and calcium (Loladze 2002). Elevated CO_2 also reduces iron and zinc concentration in edible legumes and grains (Myers et al. 2014). Likewise, higher temperature has been shown to negatively affect vitamin concentrations in horticultural crops (McKeown et al. 2006). Since the uptake of different minerals from the soil also depends on water uptake, higher CO_2 and temperature that can lead to higher VPD and reduced stomatal conductance, reduction in transpiration and water uptake may also affect mineral uptake, ultimately jeopardizing the quality of produce.

Precipitation

Climate change is expected to change the frequency, patterns and intensity of rainfall. These changes are critical for crop production in many ways including moisture stress (drought or flooding), especially if this happens during the critical stages of crop development. Particularly, the effects of drought are more important in the scenario of climate change where higher temperatures increase crop water demand. The direct effect of changes in rainfalls on crop productivity may vary, depending on the percent of agricultural areas under rainfall in different parts of the world. However, about 80% of total world cropped area is rainfed/dryland; therefore, any change in rainfall may have direct effects on global food security (Faurès et al. 2013). This is mainly because wet areas may become wetter and dry areas are expected to receive low rainfall (Liu and Allan 2013). Changes in rainfall patterns and increased variability may also create problems for individual farmers in a way that they cannot plan farming activities, as they would do under normal conditions. Changes in sowing and harvesting dates and seasonal length would ultimately result in yield loss (Linderholm 2006). Apart from the usual variations in precipitation dynamics, increased incidence of extreme weather events is another aspect of climate change that includes intensive heavy showers, flooding and hailstorms. Such events can destroy crops and may cause a delay in planting and harvesting activities, both of which can lower crop yields.

Generally, the direct effects of changes in precipitation are more relevant for areas where crops depend on seasonal rainfall or stored soil moisture. Interannual variation in rainfall patterns can cause uncertainty in farmers' minds and create problems in decision-making for farming activities. Changes in precipitation are also expected to affect crop production in irrigated areas by changing the overall hydrological cycle, mainly through its effects on the availability of freshwater in rivers, streams and groundwater reservoirs.

Interactive Effects of Climate Variables

The interactive effects of increasing temperature and elevated CO_2 are important in the context that both factors are vital for crop production and both factors can have negative and positive consequences. Elevated CO_2 is thought to have the potential to alleviate the effects of high temperature through increased photosynthesis and growth (Ainsworth and Long 2005). However, studies have indicated that, despite the fact that elevated CO_2 can mitigate the high-temperature effects on the physiology of plant processes, it may not recover the yield loss. A recent study on wheat involving elevated CO_2 and short-duration heat stress treatments during the anthesis stage concluded that elevated CO_2 mitigated the adverse effects of high temperature on photosynthesis and gas exchange; however, it does not compensate the yield loss due to direct damage of heat stress (Chavan et al. 2019). Higher crop leaf area and biomass under elevated CO_2 are also expected to increase crop water and nutrient demand. Thus, whether or not elevated CO_2 will increase crop yield depends on if the crop has access to other raw materials in a sufficient amount, especially water and nutrients. Larger canopies and higher leaf area also provide a larger surface area for transpiration loss, therefore, requiring more water.

The size and magnitude of the effects of climatic variables also depend on other non-climatic factors. Since elevated CO_2 has the potential to increase photosynthesis and crop biomass, the magnitude of this effect also depends on the availability of other raw materials, e.g. minerals (Ainsworth and Long 2005). The best example is nitrogen, where FACE experiments have indicated that the positive effects of elevated CO_2 on photosynthesis, biomass and yield are considerably lower under low than high soil nitrogen conditions (Poorter and Navas 2003; Ainsworth and Rogers 2007). Likewise, the negative effect of elevated CO_2 on the nutritional quality of plant tissues is also more pronounced under low soil N conditions (Taub and Wang 2008).

Changes in precipitation and thus the availability of water are even more important under high-temperature conditions because high temperature can be a stress itself, but, at the same time, it increases the crop water demand through increased VPD. Therefore, the two stresses at a time are definitely lethal and more difficult to manage.

Miscellaneous

Apart from changes in temperature, CO_2 and precipitation, increasing concentration of ozone and increasing threats from pest attacks are also associated with climate change. Ozone (O_3) is another environmental factor that can affect crop production by damaging the tissues, and it has been associated with climate change. Ozone is known to have negative impacts on crop yield, and its effects are likely to increase with climate change (Chuwah et al. 2015; Tai and Martin 2017). Elevated CO_2-led reduction in the stomatal opening has been shown to reduce the O_3 uptake by leaves, therefore, decreasing the exposure and minimizing its negative effects on photosynthesis, growth and yield of rice and soybean (Feng et al. 2008). Increases in temperature and humidity are likely to increase the attacks of pests such as insects and diseases. Moreover, elevated CO_2-led increase in crop growth and leaf area index would also provide suitable environments for the growth and spread of pests. The impacts of climate change on pests including diseases, insects and weeds are not well-known and require more research.

Management Strategies to Adapt Climate Change

Different management strategies have been demonstrated for adaptation of crop plants to changing climate and variability. Crop adaptations to climate change occur at different levels of agricultural association: use of quality seeds of improved varieties or species well-adopted to climate variabilities, diversification of cropping system, improved water use efficiency through irrigation and drainage systems and sustainable management of land and soil.

Development and Plantation of Plant Varieties Adapted to Climate Change

Planting material of well-adapted crop varieties is the basic requirement for high growth, quality and yields of crop plants. It is not possible to harvest a good-quality crop with low or bad quality of planting material (Gibbon 2012). Breeding efforts usually involve multilocation trials in order to develop the crop varieties that are resistant to climate-related phenomenon and more efficient to adopt the climate variabilities. For example, resistance to drought, salinity and flooding is a common climate trait for which crop varieties are bred. Other more location-specific factors include high temperature during grain filling stage, frost events at seedling or pollination stage, alternate high temperature and light rainfall which stimulate germination but obstruct seedling establishment. Increasing or maintaining the crop yield in the view of climate change largely depends on the capacity of breeders and geneticists to introduce adaptive traits found in crop plants to locally adaptive crop varieties (Jarvis et al. 2008). Active participation of farmers in the varietal development process is very important for successful adaptation of improved varieties (Efisue et al. 2008; Ashby 2009). Furthermore, introduction of resilient and adapted plant species in stress-prone areas could also be a strategy to adapt to climate change. For instance, an option includes replacing staple crops, such as maize with drought-resistant crops, such as millets and cassava. Such a shift in crop plantation can only become a viable climate-adaptive strategy, if farmers are willing to adopt new crops (Burns et al. 2010; Rezaei et al. 2015).

Conventional crop varieties are generally well-adapted to the present climatic conditions in the local crop production systems and are a potential source of adaptive genetic material for crop improvements (Mba et al. 2012; Lopes et al. 2015). However, these varieties may lose their adaptation with climate change (Bellon et al. 2011). The introduction of more suitable varieties from other places may not always be an option (Bellon and van Etten 2014). Therefore, development of new varieties through breeding seems a more viable option to adapt to climate change variations. The increased genetic vulnerability and homogeneity reduce the crop potential and make them more susceptible to the impact of climate change. This genetic vulnerability may be reduced by introducing novel traits (resistance to biotic and abiotic stresses) into the cultivars, traits often found in crop wild relatives (Lane and Jarvis 2007; Dwivedi et al. 2008). Table 1 shows some examples of the successful introduction of wild relative stress-tolerant traits into crop cultivars (Maxted and Kell 2009; Brozynska et al. 2016).

Development of climate-tolerant crop varieties required the use of a range of technologies, such as induced mutation, cell and tissue culturing, genetic engineering, marker-assisted selection and genome editing (Ahloowalia et al. 2004; Shu 2009). The development of climate-ready varieties for stress-prone areas is a key measure to deal with climate change extremes. For example, a flood-tolerant rice variety (Scuba Rice) was developed for flood-prone rice areas of Bangladesh, India and the Philippines (Singh et al. 2010). Farmers can only benefit from newly

Table 1 Examples of stress-tolerant traits obtained from wild relatives into the cultivated crop species

Cultivated crop species	Wild relatives	Traits
Rice (*Oryza sativa*)	Wild rice (*Oryza glaberrima*)	Water stress tolerance, nutritional and grain quality improvement
Oat (*Avena sativa*)	Wild oats (*Avena barbata*)	Water and heat stress tolerance
Peanut (*Arachis hypogaea*)	Wild peanuts (*Arachis cardenasii*)	Improvement in grain size, pest and disease resistance
Grape (*Vitis vinifera*)	Wild grapevine species (*Vitis amurensis*)	Cold stress tolerance in leaves
Cassava (*Manihot esculenta*)	Wild cassava (*Manihot rubricaulis*)	Adaptation to cool temperatures and high altitudes
Banana and plantain (*Musa acuminata, M. balbisiana*)	Wild plantain (*Musa balbisiana, M. nagensium*)	Drought resistance

developed climate-resistant varieties if they have timely access to the right quantity of quality seeds and planting materials. Therefore, it is important to include the effective delivery system to ensure the timely access of seeds to the farmers in the remote areas (McGuire and Sperling 2013; Westengen and Brysting 2014).

Diversification of Cropping Systems

Growing genetically diverse and improved crop varieties that are suitable for a wide range of farming practices and agroecosystems and resilient to climate variables is a valid strategy to develop the resilient crop production systems (Gibbon 2012). The level of diversification of crop species makes the difference between stressed and resilient agroecosystem. Generally, all major grain crops, for example, wheat, rice, maize, etc., are grown in monoculture systems that require significant management investment in terms of control of pests and diseases. In cropping systems, crop diversity in terms of different species and varieties is important to improve the resilience and stability of cropping systems (Folke 2006). The crop diversity also serves as an integrated pest and disease management, which has a direct impact on farm yield and revenue, since it saves a lot of external inputs and labour costs which are required for traditional management of pests and diseases. The diversity of cropping system also provides other environmental and social benefits to the society, for example, pollination and improved soil quality, and provides a wide variety of foods. Furthermore, increasing sustainable management of crop diversification will provide the food and nutritional security for the expanding urban population (Howden et al. 2007).

The diversification of cropping systems can occur at different levels, diversification of different crop varieties and species (i.e. intra- and/or inter-specific diversification), diversification at different spatial scales (i.e. individual field, farm or landscape) and diversification at different time frames. For the annual cropping system to better adapt to the climate change integration of perennial crops is a good strategy (Howden et al. 2007). Integration of perennial species in cropping system does not mean complete conversion of annual cropping system to perennial cropping or landscape dominated by perennial crops. Introduction of perennial species in annual cropping systems serves the multifunction in addition to increasing adaptability to climate change of annual crop production systems. The use of perennial crops in crop diversification gives multiple benefits, e.g. feed, food, fuel, fibre, medicines, pesticides, increased soil fertility, reduced soil erosions and serves as windbreaks. For example, integration of pigeon pea as a perennial crop with maize and soybean serves as an herbicide for weeds and provide two harvests per season (Schoeneberger et al. 2012).

At the individual farm level, the adaptability to climate change can be improved by mixing different crop varieties of the same species, for example, planting of different varieties of the same crop, which can be grown and harvested at the same time but have different responses to climate stresses (drought, heat, etc.). This strategy can successfully increase the crop adaptability to unpredictable raining season and heat; it also improves the crop stability and yields. The different management options where different varieties can intergrade together or grow one after the other to improve adaptability to climate change include relay cropping, intercropping and crop rotation (Scialabba and Müller-Lindenlauf 2010).

Improved Management of Water Resources

Sustainable management of water resources such as deficit irrigation, reduce unproductive evaporation losses and conservation measures for soil water can limit the risk of lower crop yield due to limited water. These management options are required to be adopted at different scales: at the farm field level, at the watershed or aquifer level, at the river basins level and at the national level. The farm-level adaptations are spontaneous and can perform in response to a specific change, but adaptations at other levels need advanced planning and financial support (Cooper et al. 2008; Mwongera et al. 2017).

At the farmer field scale, reduction of water losses and the soil capacity to restore rainwater can increase the resilience of cropping systems to water shortage. Different management options are adopted by farmers to improve water storage in soils, for example, on-farm water retention, on-farm water harvesting and enhanced water infiltration. These management options can be combined with efficient irrigation techniques (e.g. deficit irrigation) that was developed to reduce evaporation losses and increase crop yield per volume of water applied (Cooper et al. 2008). Selection and diversification of drought- and heat-resistant varieties will benefit farmers to cope with adverse climatic conditions. Furthermore, farmers need to be more systematic in developing drainage facilities to cope with heavy rainfalls and flood events to prevent crop damage and soil erosion (Mwongera et al. 2017). Habitat engineering and reintegration will be required to lower the impacts of flooding, provide essential nutrients to soil and control erosion.

Modern irrigation schemes are considered a step forward to adapt the climate change. Irrigation modernization requires a better mechanism for water allocation, timely alert for water scarcity for farmers, infrastructure development at local scale and management to allow more flexible and reliable water distribution (Renault et al. 2007). The establishment of the water market and water pricing are often promoted as a management tool to reduce water losses and improve water use efficiency. However, these options are difficult to implement in some places due to institutional and technical reasons. Improved weather prediction and hydrological monitoring can play a significant role in development of efficient adaptation strategies (Faurès et al. 2010). At present, weather prediction is limited to a few days. Though, better forecasting in terms of time and consistency over the season will provide the opportunity to the farming communities to better respond to climate variabilities. More efforts should be given to the timely delivery of information to farmers and increase their capacity to better utilize climate information (Gommes et al. 2010).

Sustainable Management of Land and Soil

Sustainable management of agricultural lands is important to reduce the climate change effects imposed by greenhouse gas emissions and to increase the soil carbon storage. Increasing crop productivity per unit area will eliminate the need for more land for crop production and will eventually reduce the overall emissions of greenhouse gases caused by the expansion of agricultural land. The more economical management strategies for sustainable land intensification comprise attaining the balanced nutrient cycling through protecting the soil on field and crop management (Scialabba and Müller-Lindenlauf 2010; Bitew and Abera 2018).

Soil protection can be achieved by implementing conservative tillage practices and sustainable management of crop residues in the fields. Conservation agriculture provides a strategic point for adaptation to climate change. Minimum soil disturbances, retaining the crop residues on soil and integrating different crops in rotation, reduce the soil erosion and restore the degraded soils (Farooq et al. 2011). Conservation agriculture allows to develop a more sustainable soil ecosystem and reduce the dependence on external inputs (Ghosh and Hazra 2014; Bitew and Abera 2018). Conservation tillage keep the crop residues on surface, which stabilize soil temperature, reduce moisture and nutrient losses from soil and help in the development of soil fertility. Adaptation of conservation tillage improves the growth and activities of soil microorganisms, e.g. earthworms, mites, millipedes, etc. The soil microorganisms perform the natural tillage operation, which improves the soil porosity and fertility. Organic matter accumulated by soil microorganisms improves the soil water storage capacity, which helps the crops to survive during the drought period (Lal 2004; Ghosh and Hazra 2014).

Reducing cultivation practices and retaining crop residues influence the soil carbon and nitrogen balances. Carbon is accumulated in the soil when the net nitrogen input (i.e. mineral fertilizers, organic sources, natural fixation) in the soil is greater than net removal (i.e. crop harvest, leaching, atmospheric emissions) (Corsi et al. 2012). Soil positive nitrogen balance can be attained through effective crop rotations. Keeping the soil evenly covered with crop residues with carbon to nitrogen ratio of 25–30 creates a positive residual fertilizer effect. The ideal carbon to nitrogen ratio (25–30) in crop residues can be achieved by rotating the crops high in carbon with crops high in nitrogen (Gál et al. 2007). This allows the carbon to be stored in the soil and nitrogen be released slowly to be available for the next crop (Al-Kaisi et al. 2008). Adopting the more complex crop rotation with the integration of leguminous crop increases the net carbon accumulation in soil. Crops with deep root systems accumulate carbon in the deeper soil layer which is not readily available for oxidation (Jarecki and Lal 2003).

Future Thrusts

The success of adaptation strategies to climate change depends on the participation of all key stakeholders including farmers in the development process. Interdisciplinary participatory approaches are required to develop more feasible adaptation strategies at farmers' field scale. Further, the development of targeted adaptation strategies is more important, since the adaptation strategies that are feasible at one place and for one community of farmers may not be feasible to adapt at other places. Similarly, farmers in poor and developing countries are more vulnerable to climate change. The development of adaptation strategies that are feasible for farmers under their local conditions and resources is far more important to deal with adverse impacts of climate change. The improvement in socio-economic conditions in rural areas will have a positive impact on adaptation to climate change, since it will increase the resilience of local communities to better cope the climate extremes and will reduce their vulnerability.

Technology transfer and knowledge sharing are appropriate ways to improve the adaptive capacity of resource-poor farmers. The availability of timely, accurate and easy-to-understand information about the weather forecast, pest or disease outbreaks, etc. in marginal areas will help the farmers to better prepare and plan their activities. Therefore, it is important that the information provided to the farmers are timely, up-to-date and in their local languages. In addition, availability of improved crop varieties (e.g. drought-resistant and heat-tolerant) and development of mechanization capacity of resource-poor farmers, who are more vulnerable to climate change, will increase their capacity to adapt the climate variabilities.

There is a need to advance our understanding about the interaction of two important climate variables (i.e. increasing temperature and elevated CO_2) on crop growth and quality. For example, how much of elevated CO_2 help to reduce the adverse effect associated with increasing temperature and drought stress? Similarly, how much of increasing temperature and water stress reduce the positive effect of increasing CO_2 due to direct damage to crop and decrease in crop quality (change in protein and mineral contents in grains).

References

Ahloowalia B, Maluszynski M, Nichterlein K (2004) Global impact of mutation-derived varieties. Euphytica 135:187–204
Ainsworth EA (2008) Rice production in a changing climate: a meta-analysis of responses to elevated carbon dioxide and elevated ozone concentration. Glob Chang Biol 14:1642–1650
Ainsworth EA, Long SP (2005) What have we learned from 15 years of free-air CO_2 enrichment (FACE)? A meta-analytic review of the responses of photosynthesis, canopy properties and plant production to rising CO_2. New Phytol 165:351–372
Ainsworth EA, Rogers A (2007) The response of photosynthesis and stomatal conductance to rising [CO_2]: mechanisms and environmental interactions. Plant Cell Environ 30:258–270
Alexander LV, Zhang X, Peterson TC, Caesar J, Gleason B, Klein Tank A, Haylock M, Collins D, Trewin B, Rahimzadeh F (2006) Global observed changes in daily climate extremes of temperature and precipitation. J Geophys Res Atmos:111
Al-Kaisi MM, Kruse ML, Sawyer JE (2008) Effect of nitrogen fertilizer application on growing season soil carbon dioxide emission in a corn–soybean rotation. J Environ Qual 37:325–332
Ashby JA (2009) The impact of participatory plant breeding. In: Ceccarelli S, Guimaraes EP, Weltzien E (eds) Plant breeding and farmer participation. FAO, Rome, pp 649–671
Asseng S, Ewert F, Martre P, Rötter RP, Lobell DB, Cammarano D, Kimball B, Ottman MJ, Wall G, White JW (2015) Rising temperatures reduce global wheat production. Nat Clim Chang 5:143
Bellon MR, van Etten J (2014) Climate change and on-farm conservation of crop landraces in centres of diversity. Plant Genet Resour Clim Chang:137–150
Bellon MR, Hodson D, Hellin J (2011) Assessing the vulnerability of traditional maize seed systems in Mexico to climate change. Proc Natl Acad Sci 108:13432–13437
Bitew Y, Abera M (2018) Conservation agriculture based annual intercropping system for sustainable crop production: a review. Indian J Ecol 45:235–249
Brozynska M, Furtado A, Henry RJ (2016) Genomics of crop wild relatives: expanding the gene pool for crop improvement. Plant Biotechnol J 14:1070–1085
Burns A, Gleadow R, Cliff J, Zacarias A, Cavagnaro T (2010) Cassava: the drought, war and famine crop in a changing world. Sustainability 2:3572–3607
Cape J (2008) Surface ozone concentrations and ecosystem health: past trends and a guide to future projections. Sci Total Environ 400:257–269
Chavan SG, Duursma RA, Tausz M, Ghannoum O (2019) Elevated CO_2 alleviates the negative impact of heat stress on wheat physiology but not on grain yield. J Exp Bot 70:6447
Chuwah C, van Noije T, van Vuuren DP, Stehfest E, Hazeleger W (2015) Global impacts of surface ozone changes on crop yields and land use. Atmos Environ 106:11–23
Cooper P, Dimes J, Rao K, Shapiro B, Shiferaw B, Twomlow S (2008) Coping better with current climatic variability in the rain-fed farming systems of sub-Saharan Africa: an essential first step in adapting to future climate change? Agric Ecosyst Environ 126:24–35
Corsi S, Friedrich T, Kassam A, Pisante M, Sá J (2012) Soil organic carbon accumulation and carbon budget in conservation agriculture: a review of evidence. Integr Crop Manag:16
Crafts-Brandner SJ, Salvucci ME (2000) Rubisco activase constrains the photosynthetic potential of leaves at high temperature and CO2. Proc Natl Acad Sci U S A 97:13430–13435
Dai A (2011) Drought under global warming: a review. Wiley Interdiscip Rev Clim Chang 2:45–65

De Graaff MA, Van Groenigen KJ, Six J, Hungate B, van Kessel C (2006) Interactions between plant growth and soil nutrient cycling under elevated CO2: a meta-analysis. Glob Chang Biol 12:2077–2091

Dwivedi SL, Upadhyaya HD, Stalker HT, Blair MW, Bertioli DJ, Nielen S, Ortiz R (2008) Enhancing crop gene pools with beneficial traits using wild relatives. Plant Breed Rev 30:179

Easterling DR, Wehner MF (2009) Is the climate warming or cooling? Geophys Res Lett 36

Eckersten H, Blombäck K, Kätterer T, Nyman P (2001) Modelling C, N, water and heat dynamics in winter wheat under climate change in southern Sweden. Agric Ecosyst Environ 86:221–235

Efisue A, Tongoona P, Derera J, Langyintuo A, Laing M, Ubi B (2008) Farmers' perceptions on rice varieties in Sikasso region of Mali and their implications for rice breeding. J Agron Crop Sci 194:393–400

Farooq M, Flower K, Jabran K, Wahid A, Siddique KH (2011) Crop yield and weed management in rainfed conservation agriculture. Soil Tillage Res 117:172–183

Farooq M, Hussain M, Siddique KH (2014) Drought stress in wheat during flowering and grain-filling periods. Crit Rev Plant Sci 33:331–349

Faurès J-M, Bernardi M, Gommes R (2010) There is no such thing as an average: how farmers manage uncertainty related to climate and other factors. Int J Water Resour Dev 26:523–542

Faurès J, Bartley D, Bazza M, Burke J, Hoogeveen J, Soto D, Steduto PJF (2013) Climate smart agriculture sourcebook. FAO, Rome, p 557

Feng Z, Kobayashi K, Ainsworth EA (2008) Impact of elevated ozone concentration on growth, physiology, and yield of wheat (Triticum aestivum L.): a meta-analysis. Glob Chang Biol 14:2696–2708

Field CB, Barros V, Stocker TF, Dahe Q (2012) Managing the risks of extreme events and disasters to advance climate change adaptation: special report of the Intergovernmental Panel on Climate Change. Cambridge University Press, New York

Folke C (2006) Resilience: the emergence of a perspective for social–ecological systems analyses. Glob Environ Chang 16:253–267

Gál A, Vyn TJ, Michéli E, Kladivko EJ, McFee WW (2007) Soil carbon and nitrogen accumulation with long-term no-till versus moldboard plowing overestimated with tilled-zone sampling depths. Soil Tillage Res 96:42–51

Ghosh P, Hazra K (2014) Constraints, issues and opportunities of RCT in pulse based cropping systems. Resour Conserv Technol Pulses: 18–30

Gibbon D (2012) Save and grow: a policymaker's guide to the sustainable intensification of smallholder crop production. Food and Agriculture Organization of the United Nations, Rome, p 112. US $45.00. ISBN:978-92-5-106871-7. Exp Agric 48:154–154

Gommes R, Acunzo M, Baas S, Bernardi M, Jost S, Mukhala E, Ramasamy S (2010) Communication approaches in applied agrometeorology. In: Applied agrometeorology. Springer, Berlin, pp 263–286

Hatfield JL, Boote KJ, Kimball BA, Ziska L, Izaurralde RC, Ort D, Thomson AM, Wolfe D (2011) Climate impacts on agriculture: implications for crop production. Agron J 103:351–370

Howden SM, Soussana J-F, Tubiello FN, Chhetri N, Dunlop M, Meinke H (2007) Adapting agriculture to climate change. Proc Natl Acad Sci 104:19691–19696

IPCC (2001) Climate change 2001: the scientific basis. Intergovernmental Panel on Climate Change. Cambridge University Press, Cambridge

Izaurralde RC, Rosenberg NJ, Brown RA, Thomson AM (2003) Integrated assessment of Hadley Center (HadCM2) climate-change impacts on agricultural productivity and irrigation water supply in the conterminous United States: Part II. Regional agricultural production in 2030 and 2095. Agric For Meteorol 117:97–122

Jarecki MK, Lal R (2003) Crop management for soil carbon sequestration. Crit Rev Plant Sci 22:471–502

Jarvis A, Upadhyaya HD., Gowda C, Agrawal P, Fujisaka S, Anderson B (2008) Climate change and its effect on conservation and use of plant genetic resources for food and agriculture and associated biodiversity for food security

Kjøhl M, Nielsen A, Stenseth NC (2011) Potential effects of climate change on crop pollination. Food and Agriculture Organization of the United Nations (FAO), Rome
Lal R (2004) Carbon emission from farm operations. Environ Int 30:981–990
Lal R (2005) Climate change, soil carbon dynamics, and global food security. In: Lal R, Stewart B, Uphoff N (eds) Climate change and global food security. CRC Press, Boca Raton, pp 113–143
Lane A, Jarvis A (2007) Changes in climate will modify the geography of crop suitability: agricultural biodiversity can help with adaptation
Law RD, Crafts-Brandner SJ (1999) Inhibition and acclimation of photosynthesis to heat stress is closely correlated with activation of ribulose-1, 5-bisphosphate carboxylase/oxygenase. Plant Physiol 120:173–182
Leakey AD, Ainsworth EA, Bernacchi CJ, Rogers A, Long SP, Ort DR (2009) Elevated CO2 effects on plant carbon, nitrogen, and water relations: six important lessons from FACE. J Exp Bot 60:2859–2876
Lesk C, Rowhani P, Ramankutty N (2016) Influence of extreme weather disasters on global crop production. Nature 529:84
Linderholm HW (2006) Growing season changes in the last century. Agric For Meteorol 137:1–14
Liu C, Allan RP (2013) Observed and simulated precipitation responses in wet and dry regions 1850–2100. Environ Res Lett 8:034002
Lobell DB, Gourdji SM (2012) The influence of climate change on global crop productivity. Plant Physiol 160:1686–1697
Lobell DB, Hammer GL, McLean G, Messina C, Roberts MJ, Schlenker W (2013) The critical role of extreme heat for maize production in the United States. Nat Clim Chang 3:497
Loladze I (2002) Rising atmospheric CO2 and human nutrition: toward globally imbalanced plant stoichiometry? Trends Ecol Evol 17:457–461
Long SP, Ainsworth EA, Leakey AD, Nösberger J, Ort DR (2006) Food for thought: lower-than-expected crop yield stimulation with rising CO2 concentrations. Science 312:1918–1921
Lopes MS, El-Basyoni I, Baenziger PS, Singh S, Royo C, Ozbek K, Aktas H, Ozer E, Ozdemir F, Manickavelu A (2015) Exploiting genetic diversity from landraces in wheat breeding for adaptation to climate change. J Exp Bot 66:3477–3486
Magadza CHD (2000) Climate change impacts and human settlements in Africa: prospects for adaptation. Environ Monit Assess 61:193–205
Maxted N, Kell S (2009) Establishment of a global network for the in situ conservation of crop wild relatives: status and needs. FAO Commission on Genetic Resources for Food and Agriculture, Rome, pp 266–509
Mba C, Guimaraes EP, Ghosh K (2012) Re-orienting crop improvement for the changing climatic conditions of the 21st century. Agric Food Secur 1:7
McGuire S, Sperling L (2013) Making seed systems more resilient to stress. Glob Environ Chang 23:644–653
McKeown A, Warland J, McDonald M (2006) Long-term climate and weather patterns in relation to crop yield: a minireview. Can J Bot 84:1031–1036
Monneveux P, Pastenes C, Reynolds MP (2003) Limitations to photosynthesis under light and heat stress in three high-yielding wheat genotypes. J Plant Physiol 160:657–666
Munns R, James RA, Sirault XR, Furbank RT, Jones HG (2010) New phenotyping methods for screening wheat and barley for beneficial responses to water deficit. J Exp Bot 61:3499–3507
Mwongera C, Shikuku KM, Twyman J, Läderach P, Ampaire E, Van Asten P, Twomlow S, Winowiecki LA (2017) Climate smart agriculture rapid appraisal (CSA-RA): a tool for prioritizing context-specific climate smart agriculture technologies. Agric Syst 151:192–203
Myers SS, Zanobetti A, Kloog I, Huybers P, Leakey AD, Bloom AJ, Carlisle E, Dietterich LH, Fitzgerald G, Hasegawa T (2014) Increasing CO 2 threatens human nutrition. Nature 510:139
Peters GP, Marland G, Le Quéré C, Boden T, Canadell JG, Raupach MR (2012) Rapid growth in CO2 emissions after the 2008–2009 global financial crisis. Nat Clim Chang 2:2
Poorter H, Navas ML (2003) Plant growth and competition at elevated CO2: on winners, losers and functional groups. New Phytol 157:175–198

Rashid MA, Jabloun M, Andersen MN, Zhang X, Olesen JE (2019) Climate change is expected to increase yield and water use efficiency of wheat in the North China Plain. Agric Water Manag 222:193–203

Reddy V, Pachepsky YA (2000) Predicting crop yields under climate change conditions from monthly GCM weather projections. Environ Model Softw 15:79–86

Renault D, Facon T, Wahaj R (2007) Modernizing irrigation management: the MASSCOTE approach – mapping system and services for canal operation techniques. FAO, Rome

Rezaei EE, Gaiser T, Siebert S, Ewert F (2015) Adaptation of crop production to climate change by crop substitution. Mitig Adapt Strateg Glob Chang 20:1155–1174

Rogers A, Ainsworth EA, Leakey AD (2009) Will elevated carbon dioxide concentration amplify the benefits of nitrogen fixation in legumes? Plant Physiol 151:1009–1016

Salvucci ME, Crafts-Brandner SJ (2004) Inhibition of photosynthesis by heat stress: the activation state of Rubisco as a limiting factor in photosynthesis. Physiologia Plantarum 120:179–186

Salvucci ME, Osteryoung KW, Crafts-Brandner SJ, Vierling E (2001) Exceptional sensitivity of Rubisco activase to thermal denaturation in vitro and in vivo. Plant Physiol 127:1053–1064

Sangeetha K, Thomas ASA, Seema B (2018) Comparison of organic and conventional farmers based on integrated climate change adaptive capacity. Indian Res J Ext Educ 18:1–6

Schoeneberger M, Bentrup G, De Gooijer H, Soolanayakanahally R, Sauer T, Brandle J, Zhou X, Current D (2012) Branching out: agroforestry as a climate change mitigation and adaptation tool for agriculture. J Soil Water Conserv 67:128A–136A

Scialabba NE-H, Müller-Lindenlauf M (2010) Organic agriculture and climate change. Renew Agric Food Syst 25:158–169

Sheffield J, Wood EF (2008) Global trends and variability in soil moisture and drought characteristics, 1950–2000, from observation-driven simulations of the terrestrial hydrologic cycle. J Clim 21:432–458

Shu Q (2009) Induced plant mutations in the genomics era. In: Proceedings of an international joint FAO/IAEA symposium, Vienna, Austria, 2008. Joint FAO/IAEA Division of Nuclear Techniques in Food and Agriculture

Singh US, Bari MA, Kumar A, Ismail A, Mackill DJ (2010) Status of stress tolerant rice in South Asia. In: Proceedings of the workshop on Public and Private Participation in the Development and Adoption of Stress Tolerant Rice Varieties in the coastal region of Bangladesh. Organised by the International Finance Corporation and South Asia Enterprise Development Facility, Dhaka, 19 August 2009

Tai AP, Martin MV (2017) Impacts of ozone air pollution and temperature extremes on crop yields: spatial variability, adaptation and implications for future food security. Atmos Environ 169:11–21

Taub DR, Wang X (2008) Why are nitrogen concentrations in plant tissues lower under elevated CO2? A critical examination of the hypotheses. J Integr Plant Biol 50:1365–1374

Taub DR, Miller B, Allen H (2008) Effects of elevated CO2 on the protein concentration of food crops: a meta-analysis. Glob Chang Biol 14:565–575

Westengen OT, Brysting AK (2014) Crop adaptation to climate change in the semi-arid zone in Tanzania: the role of genetic resources and seed systems. Agric Food Secur 3:3

Will RE, Wilson SM, Zou CB, Hennessey TC (2013) Increased vapor pressure deficit due to higher temperature leads to greater transpiration and faster mortality during drought for tree seedlings common to the forest–grassland ecotone. New Phytol 200:366–374

Ziska LH, Blumenthal DM, Runion GB, Hunt ER, Diaz-Soltero H (2011) Invasive species and climate change: an agronomic perspective. Clim Chang 105:13–42

Impacts of Climate Change on Weeds, Insect Pests, Plant Diseases and Crop Yields: Synthesis

Khawar Jabran, Singarayer Florentine, and Bhagirath Singh Chauhan

Introduction

Long-term weather data records have helped humans to determine that climate change is happening and will have serious impacts on the agricultural and other ecosystems. The most conspicuous of these are the warming of the earth's atmosphere and rise in its carbon dioxide (CO_2) level and changes in the availability of water. Anthropogenic activities (particularly burning of the fossil fuels) have increased (and continue to increase) the concentration of greenhouse gases in the earth atmosphere particularly the CO_2 concentrations. The current level of CO_2 in the atmosphere is more than 400 ppm. This rise in atmospheric CO_2 levels, in turn, causes warming of the global environment (Gowdy 2020). Climate change also includes difficult-to-predict changes in the rainfall patterns and amount along with the rise in the earth's surface temperature (Hawkins et al. 2020).

Not only the plants, but their pests have also been impacted by the recent climate changes. CO_2 and warming have been reported to have a contrasting effect on crop growth and yields; positive effects of CO_2 and a negative effect of the warming are on the record (Sarkar et al. 2020; Wang et al. 2020). The positive effects of high CO_2 on weeds while a change in the efficacy of the herbicides have been reported under

K. Jabran (✉)
Department of Plant Production and Technologies, Faculty of Agricultural Sciences and Technologies, Nigde Omer Halisdemir University, Nigde, Turkey

S. Florentine
Centre for Environmental Management, Federation University Australia, Mt Helen, Ballarat, VIC, Australia
e-mail: s.florentine@federation.edu.au

B. S. Chauhan
The Centre for Crop Science, Queensland Alliance for Agriculture and Food Innovation (QAAFI), The University of Queensland, Gatton, QLD, Australia
e-mail: b.chauhan@uq.edu.au

K. Jabran et al. (eds.), *Crop Protection Under Changing Climate*,
https://doi.org/10.1007/978-3-030-46111-9_8

the changing climate (Varanasi et al. 2016). Several of the insect pests and disease pathogens are also likely to receive a positive effect of climate change that will help them to increase their range and distribution and invade new areas in different geographic regions of the world.

Literature available regarding the influences of climate change on agriculture is comparatively less than the aspects of life (Callaghan et al., 2020). The book *Crop Protection Under Climate Change* is an effort to synthesize high-quality literature for understanding the effects of climate change on ecology and biology of various plant pests, their distribution in different geographical regions of the world (after an impact of climate change), likeliness of the invasiveness of pests to new areas and the control of these pests under the climate change. The book also focuses on the yield declines caused by climate change either directly or indirectly. This final chapter of the book summarizes the effects of climate change on crop plants and their pests. The chapter also summarizes how the effects of climate change could be mitigated on crop production and protection.

Impact of Climate Change on Weeds and Their Control

Increase in CO_2 is supposed to increase the growth, distribution range and invasiveness of weeds. C_3 weeds will receive more of the positive effects of the rising CO_2 than the C_4 ones, and this is likely to make C_3 weeds more competitive against crop plants (Korres et al. 2016). High CO_2 increased the biomass and growth of weeds such as *Lactuca serriola* L., *Potentilla recta* L., *Hordeum murinum* L. and *Bromus tectorum* L. (Jabran et al. 2015; Jabran and Doğan 2018). Similarly, the tiller number and biomass of *Echinochloa colona* (L.) Link. were increased if the weed was grown under 750 ppm CO_2 instead of 450 ppm (Mollaee et al. 2020). In another study, under the high CO_2 levels, the root/shoot ratio and dry weight of *Cirsium arvense* L. were increased significantly making more difficult the control of the weed (Ziska et al. 2004). Further, herbicide tolerance by weeds has also been reported to be increased under the elevated CO_2 (Ziska and Teasdale 2000). The reasons for such reduced efficacies of herbicides are the changes in biochemistries and morphologies of weed plants; these changes are resulted by the influence of environmental warming, rising atmospheric CO_2 and other factors of climate change (Ziska 2016). Although the greater tolerance to herbicides in weeds is sometimes associated with high biomass production or plant size and a subsequent dilution of the applied herbicides, there are studies that negate this mechanism (Ziska and Teasdale 2000; Manea et al. 2011). Inconsistency in results is a problem with studies that show a declined efficacy of herbicides under the changing climatic conditions (Waryszak et al. 2018). This means, contrary to most of the studies, no change in efficacy of herbicides under the simulated climatic conditions has also been reported (Jabran and Doğan 2018). A possible explanation is that the herbicide tolerance by weeds under the simulated climate varies with the species (i.e. this is species-specific). A large part of literature indicates a reduced efficacy of herbicides under the changing climate, and, subsequently, there will be difficulties in realizing good weed control. This suggests that the role of climate change in impacting the weeds and their management should be understood thoroughly and novel tools be employed for sustainable weed control.

Impact of Climate Change on Insect Pests and Their Control

Climatic changes are expected to impact both the insect pests and their natural enemies. Recent research work establishes that infestation of crops by insect pests will increase under the changing climate subsequently resulting in enhanced crop damage (Deutsch et al. 2018). For example, climate change will support *Bagrada hilaris* (an insect pest of brassica plants) to increase its distribution and invade new regions in the world (Carvajal et al. 2019). Climate change factors change biochemistries of the crop plants (e.g. changed C/N ratios) and subsequently weaken the plant defence against insect pests (Trębicki et al. 2017). This ultimately changes the feeding behaviours, population and dispersal of the insect pests (Trębicki et al. 2017). Expected global warming has been found to increase the insect pest pressure on crops; this will also support the insect pest to expand their range and invade the temperate areas that were previously too cool to inhabit those insect pests (Taylor et al. 2018).

Impact of Climate Change on Disease Pathogens and Their Control

Disease pathogens are among the most important pests of crops. Climate change either in the form of global warming, elevated CO_2 or erratic water supplies to crop plants will influence the plant disease pathogens and the crop disease incidences. For example, future climatic conditions are expected to increase the incidence of wheat diseases (e.g. *Septoria tritici* blotch, *Fusarium* head blight) in different European countries (Juroszek and von Tiedemann 2013).

It has been established that the climate change factors cause anatomical, morphological and physiological changes in plants (Watling et al. 2000; Riikonen et al. 2008). Changes in the characteristics or functioning of stomata (in response to climate change) are important in relation to disease pathogen attack on plants because stomata are the places for pathogen entry into the plant tissues (Melotto et al. 2006, 2008; Zeng et al. 2010). For instance, guard cell length was found to have a relation with the infection biology of leaf spot disease fungus (Riikonen et al. 2008). Modified leaf epidermis as a result of high CO_2 concentration increased the hostility of *Erysiphe cichoracearum* against *Arabidopsis thaliana* (Lake and Wade 2009).

Of the major impacts of climate change on plant pathogens includes the influence of warming on the prevalence and range of the pathogens; warming will enable several of plant pathogens to new areas that were previously not under invasion of those pathogens. Another fact is that the defence of host plants against the disease pathogens under climate change is likely to be disturbed.

Impact of the Climate Change Factors on Crop Yields

Global warming, rise in atmospheric CO_2 and erratic water availability have implications for the crop yields and quality (Kizildeniz et al. 2018). High CO_2 levels are supposed to increase the growth and yield of plants (with a more conspicuous effect on C_3 ones than the C_4 ones). On the other hand, the rising temperatures will have a negative effect on yield and growth of the crops. Interestingly, it is expected that rising CO_2 and temperatures will be compensating each other's effects on the crop plants. Nevertheless, growth stages of crop plants are sensitive to temperature, and physiological functions of crop plants (along with other factors) are dependent on the atmospheric CO_2 concentration and availability of moisture to plants. Recent research work implies complex and serious effects of the climate change factors on growth of crops and their yield. For example, a meta-analysis of 1700 studies showed that without any adaptations in tropical and temperate regions, the yield of rice, wheat and maize will decrease under a temperature increase of 2 °C (Challinor et al. 2014). However, the yield of these crops could be increased by 4–15% if crop level adaptations were made (Challinor et al. 2014). Another meta-analysis showed that wheat yields were decreased by greater than 50% if climate warming was higher than 2.3 °C (considering there is no change in water availability and atmospheric CO_2 levels) (Wilcox and Makowski 2014). However, a combination of elevated CO_2 and warming along with reduced precipitation by 20% could potentially increase the wheat yield (Wilcox and Makowski 2014).

It may be concluded that both negative and positive effects of climate change factors have been reported on crop yields. It is important that serious negative effects of warming on crop yield will be avoided through increased atmospheric CO_2 and, in an opposite situation, the positive effects of increased atmospheric CO_2 will be nulled through climate warming. Crop adaptations will be important in maintaining or even increasing the crop yields under climate change.

Mitigation of Adverse Impacts of Climate Change on Crop Production and Protection

Climate change brings great challenges to human interests and well-being directly and indirectly (Hobbie and Grimm 2020). Human welfare desires mitigating the adverse impacts of climate change on crop production and protection. Both the mitigation and adaptation strategies can help to deal with the adversaries of climate change (Grafakos et al. 2020). Already there have been efforts to put forward novel technologies that can aid in combating the negative effects of climate change on crops. One of the examples includes the development of crop cultivars resilient to change in climate (Ingvordsen et al. 2015; Mäkinen et al. 2015).

Crop diversity or diversity within a specific cultivated plant (cultivar diversity) can be promising in mitigating the adverse effects of climate change on growth and yield of cultivated plants. For example, diversifying the cultivars in viticulture was promising in decreasing the severe impacts of warming on the yield; this cultivar diversification could decrease the negative effects of warming (2–4 °C) by half to one-third (Morales-Castilla et al. 2020). Similarly, one of the solutions to tackle insect pests under changing climate is to maintain biodiversity. Maintaining biodiversity can help to build population of natural enemies of insect pests and hence achieve natural pest control under the changing climate.

Other than the research efforts, policymaking (at country, regional and international levels) and awareness among the farmers in the perspective of climate change are important. Combating the climate change effects will be eased if farmers are taught about the climate changes and their impacts, along with the provision of some novel technologies.

Conclusions

Most of the studies on climate change deal with either a single or a couple of factors of the climate change. The most studies on climate change factors include warming and the rising CO_2 levels. Similarly, most of the studies investigate the effects of climate change on growth of the crops and their yield. However, it will be interesting if the climate change factors such as precipitation, warming, CO_2 levels, etc. are investigated together as a system and their effects or interactions with the plant pests and their control are studied. Such studies will help better understand the real impacts of climate change on the plant pests and their control.

Climate-smart agriculture is a way forward in the wake of recent climatic changes (Middendorf et al. 2020). This is important to understand the impact of changing climate on crop plant pests, efficacy of pesticides, the pest management methods and the crop yields. Finally, policymaking and awareness in farmers about climate change are direly needed to properly cope with this great challenge.

References

Callaghan MW, Minx JC, Forster PM (2020) A topography of climate change research. Nat Clim Chang:1–6

Carvajal MA, Alaniz AJ, Núñez-Hidalgo I, González-Césped C (2019) Spatial global assessment of the pest Bagrada hilaris (Burmeister)(Heteroptera: Pentatomidae): current and future scenarios. Pest Manag Sci 75:809–820

Challinor AJ, Watson J, Lobell DB, Howden S, Smith D, Chhetri N (2014) A meta-analysis of crop yield under climate change and adaptation. Nat Clim Chang 4:287–291

Deutsch CA, Tewksbury JJ, Tigchelaar M, Battisti DS, Merrill SC, Huey RB, Naylor RL (2018) Increase in crop losses to insect pests in a warming climate. Science 361:916–919

Gowdy J (2020) Our hunter-gatherer future: climate change, agriculture and uncivilization. Futures 115:102488

Grafakos S, Viero G, Reckien D, Trigg K, Viguie V, Sudmant A, Graves C, Foley A, Heidrich O, Mirailles J (2020) Integration of mitigation and adaptation in urban climate change action plans in Europe: a systematic assessment. Renew Sust Energ Rev 121:109623

Hawkins E, Frame D, Harringto L, Joshi M, King A, Rojas M, Sutton R (2020) Observed emergence of the climate change signal: From the familiar to the unknown. Geophysical Research Letters, 47, e2019GL086259. https://doi.org/10.1029/2019GL086259

Hobbie SE, Grimm NB (2020) Nature-based approaches to managing climate change impacts in cities. Philos Trans R Soc B 375:20190124

Ingvordsen CH, Backes G, Lyngkjær MF, Peltonen-Sainio P, Jahoor A, Mikkelsen TN, Jørgensen RB (2015) Genome-wide association study of production and stability traits in barley cultivated under future climate scenarios. Mol Breed 35:84

Jabran K, Doğan MN (2018) High carbon dioxide concentration and elevated temperature impact the growth of weeds but do not change the efficacy of glyphosate. Pest Manag Sci 74:766–771

Jabran K, Dogan MN, Eren Ö (2015) Effect of ambient and simulated CO2 on the growth invasive weed Potentilla recta L. Poljoprivreda i Sumarstvo 61:107

Juroszek P, von Tiedemann A (2013) Climate change and potential future risks through wheat diseases: a review. Eur J Plant Pathol 136:21–33

Kizildeniz T, Pascual I, Irigoyen J, Morales F (2018) Using fruit-bearing cuttings of grapevine and temperature gradient greenhouses to evaluate effects of climate change (elevated CO2 and temperature, and water deficit) on the cv. Red and white Tempranillo. Yield and must quality in three consecutive growing seasons (2013–2015). Agric Water Manag 202:299–310

Korres NE, Norsworthy JK, Tehranchian P, Gitsopoulos TK, Loka DA, Oosterhuis DM, Gealy DR, Moss SR, Burgos NR, Miller MR (2016) Cultivars to face climate change effects on crops and weeds: a review. Agron Sustain Dev 36:12

Lake JA, Wade RN (2009) Plant–pathogen interactions and elevated CO2: morphological changes in favour of pathogens. J Exp Bot 60:3123–3131

Mäkinen H, Kaseva J, Virkajärvi P, Kahiluoto H (2015) Managing resilience of forage crops to climate change through response diversity. Field Crop Res 183:23–30

Manea A, Leishman M, Downey P (2011) Exotic C 4 grasses have increased tolerance to glyphosate under elevated carbon dioxide. Weed Sci 59:28–36

Melotto M, Underwood W, Koczan J, Nomura K, He SY (2006) Plant stomata function in innate immunity against bacterial invasion. Cell 126:969–980

Melotto M, Underwood W, He SY (2008) Role of stomata in plant innate immunity and foliar bacterial diseases. Annu Rev Phytopathol 46:101–122

Middendorf BJ, Prasad PV, Pierzynski GM (2020) Setting research priorities for tackling climate change. J Exp Bot 71:480–489

Mollaee M, Mobli, Chauhan BS (2020) The response of glyphosate-resistant and glyphosate-susceptible biotypes of Echinochloa colona to carbon dioxide, soil moisture and glyphosate. Sci Rep 10

Morales-Castilla I, de Cortázar-Atauri IG, Cook BI, Lacombe T, Parker A, van Leeuwen C, Nicholas KA, Wolkovich EM (2020) Diversity buffers winegrowing regions from climate change losses. Proc Natl Acad Sci

Riikonen J, Syrjälä L, Tulva I, Mänd P, Oksanen E, Poteri M, Vapaavuori E (2008) Stomatal characteristics and infection biology of Pyrenopeziza betulicola in Betula pendula trees grown under elevated CO2 and O3. Environ Pollut 156:536–543

Sarkar MSK, Begum RA, Pereira JJ (2020) Impacts of climate change on oil palm production in Malaysia. Environ Sci Pollut Res:1–11

Taylor R, Herms DA, Cardina J, Moore RH (2018) Climate change and pest management: unanticipated consequences of trophic dislocation. Agronomy 8:7

Trębicki P, Dáder B, Vassiliadis S, Fereres A (2017) Insect–plant–pathogen interactions as shaped by future climate: effects on biology, distribution, and implications for agriculture. Insect Sci 24:975–989

Varanasi A, Prasad PV, Jugulam M (2016) Impact of climate change factors on weeds and herbicide efficacy. Adv Agron 107–146

Wang B, Cai W, Li J, Wan Y, Guo C, Wilkes A, You S, Qin X, Gao Q, Liu K (2020) Leaf photosynthesis and stomatal conductance acclimate to elevated [CO2] and temperature thus increasing dry matter productivity in a double rice cropping system. Field Crop Res 248:107735

Waryszak P, Lenz TI, Leishman MR, Downey PO (2018) Herbicide effectiveness in controlling invasive plants under elevated CO2: sufficient evidence to rethink weeds management. J Environ Manag 226:400–407

Watling JR, Press MC, Quick WP (2000) Elevated CO2 induces biochemical and ultrastructural changes in leaves of the C4 cereal sorghum. Plant Physiol 123:1143–1152

Wilcox J, Makowski D (2014) A meta-analysis of the predicted effects of climate change on wheat yields using simulation studies. Field Crop Res 156:180–190

Zeng W, Melotto M, He SY (2010) Plant stomata: a checkpoint of host immunity and pathogen virulence. Curr Opin Biotechnol 21:599–603

Ziska LH (2016) The role of climate change and increasing atmospheric carbon dioxide on weed management: herbicide efficacy. Agric Ecosyst Environ 231:304–309

Ziska LH, Teasdale JR (2000) Sustained growth and increased tolerance to glyphosate observed in a C3 perennial weed, quackgrass (Elytrigia repens), grown at elevated carbon dioxide. Funct Plant Biol 27:159–166

Ziska LH, Faulkner S, Lydon J (2004) Changes in biomass and root: shoot ratio of field-grown Canada thistle (Cirsium arvense), a noxious, invasive weed, with elevated CO 2: implications for control with glyphosate. Weed Sci 52:584–588

Index

A
Abiotic stresses, 146
Abutilon theophrasti, 24
Adenosine triphosphate (ATP), 149
Agricultural industries, 2
Agricultural management practices, 160
Agricultural practice, 1, 5
Agrobacterium, 89
Agroecological field-based practices, 132
Agroecosystems
  arthropod herbivores, 112
  biodiversity management, 132
  chemical insecticides, 131
  ecology, 101
  functional plants planting, 132
  predators and parasitoids, 130
Agronomic crop weeds, 11
Agronomic tools
  AMF, 156
  crop production systems, 154
  crop variety and selection, 153
  cultural practices, 152
  fallowing, 154
  irrigation, 155, 156
  mulching and stubble tillage, 154
  nutrient management, 155
  PGR, 156
  planting time, 153
  stand density, 153
  tillage practices, 153
Air pollutants, 172
Alopecurus myosuroides, 72
Amaranthus retroflexus, 24, 62
Aphid species, 112
Aphidius ervi, 125
Arable ecosystems, 61–68
Arbuscular mycorrhizal fungi (AMF), 152, 156
Arthropod herbivores
  ambient temperatures, 108, 109
  behaviors, 108
  categories, 109
  climate effects (*see* Climate direct effects, arthropod herbivores)
  climate indirect effects (*see* Arthropod natural enemies)
  correlation techniques, 109
  ectotherms, 108
  independent factors, 109
  indirect crop management-mediated climate effects, 126–130
  indirect plant-mediated climate effects (*see* Crop-herbivore interactions)
  insect populations, 109
  key attributes, 110
  population changes, 109
Arthropod migration, 114–116
Arthropod natural enemies
  crop-aphid-parasitoid systems, 122
  in situ responses, 123–126
  range expansion and voltinism, 122, 123
  variety, 121
Arthropod range expansion
  crop pests, 111
  EXAMINE network, 113
  natural enemies, 112
  phenological matching, 111
  poleward, 111
  predictive models, 112
  simulation models, 113
  temperature, 112
  tropical montane, 113

K. Jabran et al. (eds.), *Crop Protection Under Changing Climate*,
https://doi.org/10.1007/978-3-030-46111-9

Aspergillus, 94
Atmospheric CO_2
  C_3 and C_4 plants, 7
  concentrations, 104, 172
  effects, 3
  industrialization, 3
  multifaceted situation, 6
  nutrient availability management, 8
  oceanic temperatures, 4
  weather conditions, 2
Austrodanthonia caespitosa, 7

B
Bacterial pathogens, 88
Bagrada hilaris, 191
Biochar, 131
Biological weed control methods, 75
Blumeria graminis, 87
Botrytis cinerea, 88
Brassicaceae family, 23
Breeding, 179
Brown planthopper, 113–115, 129
Burkholderia andropogonis, 88

C
C_3 and C_4 plants, 7, 8
C_3 weed species, 8
Carbon, 183
Carbon dioxide (CO_2), 85, 87, 89
  atmospheric, 174
  C_3 and C_4 photosynthetic pathway, 175
  effects, 189
  elevation, 174, 175
  stomatal regulation, 175
  water availability, 189
Carbon-rich environment, 31
Carbon sequestration, 45
Changing planting date, 93
Chilo partellus, 123
Cirsium arvense, 34
Cladosporium fulvum, 90
Climate change
  agricultural practices, 145
  agricultural pressures, 18
  agriculture, 20, 21, 190
    practices, 145
    pressures, 18
  anthropogenic causes, 19
  anthropogenic emissions, 85
  atmospheric CO_2, 3, 4, 17, 172
  C_3 and C_4 plants, 7, 8
  climatic variables, 19
  CO_2 emissions, 20
  crop management strategies, 18
  crop production, 2, 171
  definition, 2
  difficult-to-predict, 189
  disease epidemics, 85
  earth's temperatures, 85
  effects, 5–6
  endemic bacterial diseases, 85–86
  energy effect, 19
  extreme weather conditions, 20
  extreme weather fluctuations, 85
  factors, 171
  foci, 2
  fungal pathogens, 87, 88
  GHGs, 19, 145
  growth mechanisms, 18
  heat-related impacts, 172
  herbicide tolerance issues, 8
  host-pathogen interactions, 86
  impact, 22–24
  insects impact (*see* Insects pest impact, CO_2 levels)
  interactions, 85
  long-term weather data, 189
  moisture stress, 20
  negative impacts, 17
  pathogens, 86
  pest insect activities, 12
  photosynthesis, 19
  physical and biochemical manifestations, 3
  plants reaction, 4, 5
  predictive modelling, 8
  surface temperatures, 19
  temperature, 86, 172
  temporal and geographic scale, 20
  virtual conviction, 171
  warming, 193
  weed control strategies, 17
  weed infestation, 21
  weed species, 17
Climate direct effects, arthropod herbivores
  arthropod range expansion, 111–114
  CO_2 and radiation, 110
  endosymbiotic microorganisms, 110
  genetic variability and genotypic flexibility, 110
  migration, 114–116
  voltinism, 114
Climate extremes, 145, 147, 151, 160, 163
Climate-proofed rainfed cropping systems, 160–161

Climate-resistant varieties, 180
Climate-smart agriculture, 147, 193
Climate variables, 145
  CO_2, 174–176
  interactive effects, 177, 178
  O_3, 178
  precipitation, 176, 177
  temperature, 173–174
Climate warming, 10
Climatic variability, 127
Climatic variables, 172
CLIMEX, 8
Cloud trend patterns, 105, 106
Cnaphalocrocis medinalis, 108
CO_2 fixation mechanism, 7
Conservation agriculture (CA), 63, 182
Conservation tillage, 182
Cooler canopies, 175
Corn leaf aphids, 126
Cotesia flavipes, 123
Crassulacean acid metabolism (CAM), 29
Critical period for weed control (CPWC), 71
Crop adaptations, 192
Crop-aphid-parasitoid systems, 122
Crop-herbivore interactions
  categories, 116
  host-plant
    resistance, 117–119
    tolerance, 120
    vulnerability, 121
Crop management practices, 162, 163
Crop management, future climates
  agroecosystems, 132
  biochar, 131
  crop managers, 130
  crop resilience, 132
  food-web complexity, 132
  functional plants, 132
  ground cover, 131
  host-plant resistance, 131
  insecticides, 131
  natural enemies, 131
  pest management practices, 130
  pest population pressure, 130
  SRI, 131
Crop managers, 130
Cropping systems
  acclimation, 60, 61
  climate change, 58, 59
  developing economies, 57
  economic losses, 57
  economies, 58
  evolution, 61
  food production, 57
  HE, 63–68
  impact of weeds, 57
  invasive weed species, 57
  migration, 60
  production, 58
  weed control practices, 57
  weed invasion, 63
  weeds and crop plants, 61, 62
Cropping systems diversification
  climate stresses, 181
  crop varieties, 180
  levels, 180
  perennial species integration, 181
  pest and disease management, 180
Crop production management
  breeding efforts, 44
  cause allergic and toxic reactions, 45
  chemical weed control, 45
  conservation agriculture systems, 44, 45
  world's food requirements, 1
Crop productivity and quality
  management strategies, 178
  plant varieties plantation, 179
  sustainable management
    land and soil, 182–183
    water resources, 181–182
  systems diversification, 180–181
  variables (*see* Climate variables)
Crop protection genetic tools
  biotechnological approach, 159
  drought tolerance, 157
  genomics, 159
  genotypes and environment, 158
  grafting, 159
  QTLs, 159
  and research, 157–158
  traditional breeding/MAB, 159
  water requirement, 156
  yield-versus-drought stress curves, 157
Crop rotations, 154
Crop variety and selection, 153
Crop-weed competition
  climate change, 37
  CO_2 enrichment, 39, 40
  global warming conditions, 38
  photosynthetic pathways, 37
  universal crop/weed axiom, 37
  variable nitrogen supply, 39
  vegetative biomass, 37
  water stress conditions, 38

Crop weed management
agricultural robotics, 73, 74
biological weed control methods, 75
competition, 70, 71
competitive cultivars, 68, 69
diverse cropping systems, 69, 70
herbicides, 75, 76
HWSC, 72, 73
IWM, 77
nanotechnology, 74
tillage, 71, 72
tools, 76
Cultural practices, 152

D
Drought adaptation, 161
Drought mitigation, 160
Drought preparedness, 160, 161
Drought-resilient agriculture
capacity, 162
crop management, 162
mitigation, 161
protected cultivation, 162
traditional farming, 162
Droughts
agronomic (*see* Agronomic tools)
climate change-induced agricultural, 163
genetic tools (*see* Crop protection genetic tools)
management practices, 147
mitigation and crop management, 160
morphology, 146
resistance, 149
stress, 146
Droughts tolerance, 146
biotechnology approach, 159
crop improvement, 157, 158
genetic transformation, 147, 159
interactions, 146
multigene trait, 157
PGR, 156
water-uptake, 159
Dry bean (*Phaseolus vulgaris* L.), 58
Dust, 106

E
Echinochloa glabrescens, 62
Ecosystem, 91, 92
Ecosystem services, 101
Effector-triggered immunity (ETI), 88, 90
Effects of drought
crop morphology
alterations, 149
environmental stress, 149
poor growth, 150
reductions, 150
water deficit, 150, 151
water stress, 150
crop physiology
CO_2 levels reduction, 149
critical water deficiency, 148
photosynthesis reduction, 148
soil moisture reduction, 148
substantial decline, 147
turgor, 147
water balance, 149
weak growth and poor stand establishment, 148
crop yields
area-weighted, 152
extreme climate event, 151
food demand, 151
food production, 151
water resources, 151
water stress, 151
Effects of warming, 31, 32
Elevated CO_2
concentration, 119
cooler canopies, 175
FACE, 175
fertilization-like effects, 175
impacts, 172
legume crops, 175
nitrogen fertilizer, 118
photosynthates, 176
quality and chemical composition, 176
stomatal conductance, 175
water use efficiency, 175
Endophytes, 131
Endosymbiotic microorganisms, 110
Environmental stress conditions, 6
Evapotranspiration (ET), 174
EXAMINE network, 113, 114
Extracellular polysaccharides (EPS), 88

F
Fallowing, 154
Food and Agriculture Organization (FAO), 157
Food-borne contamination, 94
Food safety, 94
Free-air CO_2 enrichment experiments (FACE), 7, 30, 175, 176, 178
Functional plants, 132

Fungal pathogens, 86–88
Fungicides, 93

G
General circulation models (GCM), 160
Global climate change
arthropod herbivory, 108–110
cloud cover, 105
CO_2 concentration, 102
complex global changes, 101
complex interactions, 101
decomposers/pollinators, 102
dust, 105
impact, crop yield potential, 107, 108
land-use change, 106
positive feedback systems, 105
precipitation, 105
temperature, 104
typhoons and hurricanes, 106, 107
volcanic activity, 102
Global food consumption, 102, 103
Global food security, 151, 177
Global temperature anomalies, 104, 105
Global warming
estimation, 104
migration, 114
Grafting, 159
Grain crops, 2
Greenhouse gases (GHGs), 19
Ground cover, 131

H
Harvest weed seed control (HWSC), 72, 73
Helicoverpa armigera, 118
Herbicide efficacy (HE), 63–68
Herbicide Resistance Action Committee (HRAC), 25–26
Herbicides, 75, 76
carbohydrate production, 27
chemical weed management, 24
elevated CO_2, 28, 29
global warming, 28
implications, 27
physiological processes, 24
plant functions, 24
plant processes, 24
resistance, 27, 28, 41, 46
tolerance, 190
Host-pest synchronization, 11
Host-plant ontogeny, 126
Host-plant resistance
antixenotic components, 117
atmospheric CO_2 concentrations, 117
climate change, 119
cold weather, 117
comparative experiments, 117
dust, 117, 119
elevated CO_2 concentrations, 118
herbivore fitness, 119
salt spray, 119
temperature, 117
Host-plant tolerance, 120
Host-plant vulnerability, 121
Hymenopteran parasitoids, 124

I
Impact of climate change
crop production and protection, 192, 193
crop yields, 192
disease pathogens and control, 191, 192
insect pest and control, 191
weeds and control, 190, 191
Imperata cylindrica, 32
Indirect crop management-mediated climate effects
agrochemicals, 130
crop duration, 126
factors, 127
herbivore pests, 128
hormesis, 130
insecticides, 128, 130
landscape structure changes, 128
nutrient management, 126
outbreaks, 129, 130
parallel responses, 128
pest mortality, 128
production, 127
spatial landscape complexity, 128
Insect herbivores, 101
Insects pest impact, CO_2 levels
agricultural production, 10
agronomic crop weeds, 11
Australian agriculture, 9
biological control organisms, 9
geographical range, 10
human-driven indicators, 9
issues, 11
overwinter survival, 9
plant chemical composition, 10
plant leaf epidermis, 11
seasonal temperatures, 9
soil dwelling, 10
temperature and rainfall, 9

In situ responses, natural enemies
A. ervi, 125
ant-aphid mutualism, 125
braconid parasitoid, 125
efficiency, 124
herbivores, 124, 125
hymenopteran parasitoids, 124
ladybeetle, 125
parasitism, 125
predator and prey generations, 125
S. avenae, 124
symbiont-bearing clones, 125
synchrony, 123
Integrated weed management (IWM), 68, 77
Interannual variation, 177
Interdisciplinary participatory approaches, 183
Intergovernmental Panel on Climate Change (IPCC), 19, 85, 102, 145
Intra-seasonal variability, 163
Invasive weeds, 6, 22, 23, 34
Irrigation, 155, 156

K
Koinobiont parasitoids, 123

L
Land and soil sustainable management, 182, 183
Landscape-based practices, 132
Late-sown maize, 121
Leaf water potential (lwp), 148
Legume crops, 175

M
Magnaporthe grisea, 87
Maize, 102, 107, 108
analysis, 108
elevated CO_2, 118
experiment, 120
farmlands, 112
high light intensity, 119
warm winter crops, 127
Management-induced outbreaks, 129
Modern irrigation schemes, 182
Modes of action (MOA), 75
Modified leaf epidermis, 191
Moisture stress, 150
Monophagous/oligophagous species, 128
Monosporascus cannonballus, 87
Mulching, 154
Mycotoxins, 94

N
Nanotechnology, 74
National Climate Assessment (NCA), 145
Natural resource system, 1
Northern Hemisphere land masses, 6
Northward range expansion, 111
Novel biochemical and physiological conditions, 6
Nutrient availability, 6
Nutrient management, 155

O
Ozone (O_3), 178

P
PAMP-triggered immunity (PTI), 88
Parthenium hysterophorus, 29, 63
Pathogen-associated molecular patterns (PAMPs), 88
Pepino mosaic virus (PepMV), 89
Perennial species integration, 181
Pest incidences, 101
Pest insect activities, 12
Pest management practices, 130, 131
Pest species expansion, 11
Phoma stem canker (*Leptosphaeria maculans*), 91
Photosynthesis, 173
Photosynthetically active radiation (PAR), 70
Physiological attributes
carbon fixation mechanism, 30
carboxylation kinetics, 29
complexity, 30
crop-weed competition, 30
effects of warming, 31, 32
net assimilation rate, 30
nutrient variations, 31
oxygenation reaction, 29
photosynthetic pathway, 29
plant growth reduction, 30
stomatal conductance and transpiration rate, 30
water stress, 32, 33
Physiological parameters, 148
Phytophthora infestans, 92
Plant growth, 150
Plant growth regulators (PGR), 156
Plant-herbivore interactions, 105, 107, 119
Planthoppers
brown, 121
fecundity, 118

migration, 115, 116
rice, 109
white-backed, 115, 131
Plant pathogens, 192
agroecosystems, 86
atmosphere, 87
climate change, 86, 88, 94, 95
ecosystem, 91, 92
environmental conditions, 86
environmental factors, 90
EPS, 88
ETI, 90
food safety, 94
loss of function, 90
natural climatic barriers, 89
powdery mildew and foliar blights, 87
temperatures, 87, 88, 90
Policymaking, 193
Polyculture/multiple crop production systems, 154
Polyembryony, 122
Potato leaf roll virus (PLRV), 89
Potato yellow vein virus (PYVV), 89
Precipitation, 176, 177, 193
Predictive modelling, 8
Protected cultivation, 162
Puccinia substriata, 87

Q
Quantitative trait loci (QTLs), 159
Queensland fruit fly, 10

R
Ralstonia solanacearum, 88
Range expansion, 113, 116
Range shift, 23
Representative concentration pathways (RCPs), 102
R-gene-mediated disease control, 90
Rhamphicarpa fistulosa, 37
Rice
C. medinalis, 108
elevated CO_2, 117
fields, temperature, 124
global production, 108
M. separata, 128
mortality, 124
natural enemies, 122
pests, 129
planthoppers, 115, 116, 118, 121
SRI, 131
tropical and subtropical, 102
variety IR36, 117
Rice waika virus, 121
RNA interference (RNAi) technology, 76
Rottboellia cochinchinensis, 36
Rottboellia exaltata, 23

S
Salt spray, 119
Septoria lycopersici, 87
Septoria tritici blotch (STB), 93
Sitobion avenae, 124
Socio-economic conditions, 183
Soil microorganisms, 182
Soil moisture, 172
Soil positive nitrogen balance, 183
Soil protection, 182
Stand density, 153
Stem rot (*Sclerotium rolfsii*), 93
Stomatal conductance, 7
Stress-tolerant traits, 180
Strip cropping, 154
Stubble tillage, 155
Sturmia bella, 123
Sustainable agricultural management practices, 147
Sustainable crop situation, 2
Symbiont-bearing clones, 125
System of rice intensification (SRI), 131

T
Technology transfer and knowledge sharing, 183
Temperature, 17, 19–23, 27–29, 32, 43, 44, 46
biomass and yields, 174
changes, 58
climate change-induced changes, 173
effects, 173
ET, 174
and moisture availability, 6
photosynthesis and respiration, 173
VPD, 174
warming, 174
Tillage practices, 153
Toxicodendron radicans, 35
Traditional 'tropical' weeds, 5
Tropospheric O_3 concentrations, 172
Turgor, 147

U
United Nations Environment Programme, 19

V
Vapour pressure deficit (VPD), 174
Vectors, 89
Viral pathogens, 89
Voltinism, 114

W
Warming, 193
Water deficit, 150
Water resources, 151
Water resources sustainable management, 181, 182
Water stress, 32, 33, 146, 150
Water use efficiency (WUE), 148, 174, 175, 178, 182
Weed growth behaviour
  agricultural production system, 34
  agroecosystem, 33
  complications, 34
  effects of warming, 35, 36
  environmental stresses, 33
  genetic diversity, 33
  invasive weed species, 34
  plant biomass, 33
  plant development phases, 33
  plant growth, 34
  reproductive capacity, 34
  seed yield, 34
  water stress, 36
  weed biomass, 34
Weed management
  ambient condition warming, 43
  biological and cultural methods, 41
  chemical weed control, 40
  climate change, 40
  climate variability, 40
  climatic factors, 40
  CO_2 fertilization, 41–43
  crop phytotoxicity, 40
  genetic diversity, 41
  phenotypic plasticity, 41
  water stress, 43
Wheat
  bulb fly, 121
  feeding, 125
  flowering resistance, 117
  H. armigera, 118
  outbreaks and damage, 111
  parasitoids, 124
  production, 108
  stem sawfly, 119
  temperate crop, 102
World Meteorological Organization, 19

X
Xanthium strumarium, 37

Y
Yeast-like endosymbionts, 129

The manufacturer's authorised representative in the EU is Springer Nature Customer Service Centre GmbH, Europaplatz 3, 69115 Heidelberg, Germany. If you have any concerns regarding our products, please contact ProductSafety@springernature.com

Printed and bound by CPI Group (UK) Ltd, Croydon, CR0 4YY
22/09/2026
02221060-0003